The Global Smart City

The Global Smart City: Challenges and Opportunities in the Digital Age

BY

FILIPPO MARCHESANI

University G. d'Annunzio, Italy

emerald PUBLISHING

United Kingdom – North America – Japan – India – Malaysia – China

Emerald Publishing Limited
Emerald Publishing, Floor 5, Northspring, 21-23 Wellington Street, Leeds LS1 4DL

First edition 2024

British Library Cataloguing in Publication Data
A catalogue record for this book is available from the British Library

ISBN: 978-1-83797-576-1 (Print)
ISBN: 978-1-83797-575-4 (Online)
ISBN: 978-1-83797-577-8 (Epub)

Printed and bound by CPI Group (UK) Ltd, Croydon, CR0 4YY

Contents

About the Author

Filippo Marchesani is Researcher at the Department of Management and Business Administration at the University "G. d'Annunzio" Chieti-Pescara. He holds a PhD in Accounting, Management, and Business Economics. He was a visiting researcher at the University of Girona (Spain). His research interests focus on cities competitiveness, smart cities, smart tourism, innovation and entrepreneurship. He sits on the editorial board of Springer Nature *Business & Economics Journal* and is a peer-reviewer for several international journals. He has recently published these issues in *International Marketing Review, Technological Forecasting and Social Change, Cities, Journal of Cleaner Production, Current Issues in Tourism, International Journal of Tourism Cities* and has contributed to several books.

Introduction and Overview

In a world shaped by the juxtaposing forces of digital technologies and globalization, there is an urgent need for profound reflection on the intricate interaction between contemporary cities, the urban environment, users, stakeholders, and their transformative impact on society at large. This book aims to present a comprehensive exploration of the current state of cities, focusing on both organizational and strategic perspectives, in order to better understand the complex and still understudied ecosystem of smart cities.

Drawing upon analysis of factual data, empirical evidence from primary and secondary sources, interviews, and compelling case studies, this book delves deeply into the concept of smart cities. It examines their multifaceted role in shaping our lives, critically assessing their benefits and limitations and providing implications and suggestions for various stakeholders, including public managers, policymakers, and citizens.

As smart cities assume an increasingly pivotal role in society, this book contributes to the ongoing discourse by providing novel insights into the trajectories of smart city projects. It investigates the impact of these initiatives within organizations and beyond, while also addressing the managerial challenges and consequences within the urban environment. By presenting a synthesis of theoretical lenses from diverse fields such as urban studies, management, and economics, this book offers a comprehensive understanding of the smart city phenomenon and its far-reaching implications.

This book is structured into two parts, each offering a distinct perspective to contribute to the debate: the organizational aspect in the first part and the strategic aspect in the second. This structure aims to provide a comprehensive and global understanding of the phenomenon of smart cities.

After the introductory chapter, which provides an overview and guidelines on smart cities, the first part of this book, titled "Internal Implementation of Smart City Ecosystems," delves deep into the internal implementation of smart city projects. Specifically, it examines the digital implementation, internal dimensions, and the geographical and strategic differences that influence the implementation of smart city projects.

In the first part of this book, the second chapter titled "Digital Implementation in the Smart City Ecosystem" provides an overview of the digital implementation in smart city ecosystems. It explores the role of data ecosystems and the Data Value Chain in smart cities, along with the challenges and opportunities they bring. This chapter also highlights the interaction between local and national data ecosystems and the importance of digitalization in smart city processes and services. Interviews with smart city managers, policymakers, and technicians offer valuable insights into the practical implementation and impact of data management in urban development. The third chapter, "Orchestrating the Implementation of the Smart City," focuses on the six strategic dimensions of smart city projects: governance, environment, people, living, mobility, and economy. Through in-depth analysis, case studies, and interviews, this chapter aims to provide a comprehensive understanding of these dimensions and their significance in smart city development. It challenges prevailing media portrayals of smart city strategies and explores the internal implementation and objectives of smart cities within each dimension. Chapter 4, "Geographic Patterns in Smart City Implementation," examines the geographical differences and approaches influencing smart city projects worldwide. It discusses the impact of contextual factors on smart city initiatives and explores the strategic and geographical approaches shaping their implementation. This chapter analyzes citizen-based, technology-based, and decision-making-based development in smart cities and examines the influences of different regions on smart city projects, including Europe, North America, Latin America, and East and South Asia.

After analyzing the internal structure, organizational aspects, role and management of technology, as well as the geographical and governance approaches to smart city projects, the second part of this book shifts focus to the urban, economic, and competitive outcomes of these projects. It discusses the implications of smart projects and their potential impact on various outcomes, such as internal development, economic environment, and city attractiveness. Thus, moving to the second part of this book, Chapter 5, "Urban Environment in the Smart City," delves into the urban outcomes of smart city projects. It focuses on the digital and environmental aspects of smart cities, discussing the advancements in digital systems, environmental benefits, and improved quality of life for citizens. This chapter incorporates interviews, case studies, and international research to offer a comprehensive understanding of the impact of smart cities on the urban ecosystem. The sixth chapter explores the economic and business environment of smart city projects. It highlights the potential of smart cities to foster innovation, entrepreneurship, and economic growth. By analyzing the interplay between smart cities and the entrepreneurial ecosystem, this chapter examines the economic outcomes of smart cities, including business creation, startups, and the overall economic environment. It provides insights into the relationship between smart cities and economic prosperity. Chapter 7 focuses on the competitive outcomes of smart cities. It explores how smart cities can gain a competitive edge by offering advanced and innovative environments that attract citizens, businesses, and tourists. This chapter discusses marketing innovation, internationalization, and the competitive landscape of smart cities. It emphasizes the importance of city attractiveness in a globalized world and presents challenges and future prospects for smart city development.

Concluding the book, Chapter 8 offers practical guidelines for navigating the landscape of smart cities. It synthesizes the key points covered throughout this book and addresses the challenges and opportunities of smart city initiatives. This chapter emphasizes the need for ongoing adaptation and innovation in the rapidly evolving digital age.

In conclusion, "The Global Smart City: Challenges and Opportunities in the Digital Age" aims to provide a comprehensive understanding of the smart city phenomenon and its implications. It offers valuable insights and recommendations for policymakers, public managers, and citizens. By incorporating interdisciplinary perspectives and empirical evidence, this book makes a significant contribution to the ongoing discourse on smart cities. It encourages readers to embrace the challenges and seize the opportunities presented by smart city initiatives to create sustainable, efficient, and equitable urban environments.

Chapter 1

Introducing and Understanding Smart Cities

Abstract

This chapter examines the transition from traditional cities to smart cities, with a focus on integrating digital technologies, data, and services into the urban environment. The concept of a smart city aims to create urban spaces that are more efficient, sustainable, and livable by fostering collaboration among stakeholders such as government, businesses, and citizens. This chapter highlights that this transformation encompasses not only technological advancements but also significant social, cultural, and economic implications. It emphasizes the need to rethink urban governance, citizen engagement, and public service delivery. By leveraging data for decision-making, smart city technologies enable cities to optimize urban services and streamline internal operations. Strategically, smart city development aims to attract businesses, investors, and visitors through improved services, sustainability, and accessibility. This chapter also explores how smart cities can enhance their competitiveness by promoting innovation, entrepreneurship, and a dynamic environment for residents and tourists. Moreover, it discusses how digital technologies enable smart cities to offer personalized and interactive experiences to visitors. Overall, this chapter provides a comprehensive perspective on the potential benefits and opportunities arising from smart city transformations, encompassing efficiency, sustainability, economic growth, and enhanced visitor experiences. It serves as a valuable resource for interpreting and understanding various aspects of the book.

Keywords: Smart cities; city; digital technologies; urban transformation; entrepreneurial urbanism; urban planning; smart governance

1.1 From City to Smart City

In recent years, the concept of a smart city has gained significant attention as cities around the world are transforming to adapt to the digital age. The rise of

The Global Smart City, 1–12

doi:10.1108/978-1-83797-575-420231001

digital technologies has revolutionized the way we live and work, and cities have been at the forefront of this transformation. Cities have always been centers of innovation, but the digital era has brought about new opportunities and challenges for urban development. As a result, cities are now transforming into "smart cities" with the aim of creating more efficient, sustainable, and livable urban environments (Albino, Berardi, & Dangelico, 2015; Caragliu, del Bo, & Nijkamp, 2011; Vanolo, 2014).

Throughout this book, we will delve deeper into these dynamics and conjectures to provide a broader understanding of the transformation from a traditional city to a smart city. We will analyze the integration of digital technologies, data, and services into the urban environment, aiming to create more efficient, sustainable, and livable urban spaces as well as the outcomes and the effective impact of those transformations. However, it is crucial to note that this change goes beyond the urban realm and carries significant social, cultural, and economic implications. It necessitates a fundamental shift in the internal and external management and operation of cities, calling for a re-imagination of urban governance, citizen engagement, and public service delivery.

The smart city approach places great emphasis on collaboration among various stakeholders, including government, businesses, and citizens, in order to co-create innovative solutions and tackle complex urban challenges. By embracing the concept of smart cities, cities aspire not only to enhance their infrastructure and service delivery but also to foster social inclusion, technological adoption, and economic growth. The integration of digital technologies enables cities to improve residents' quality of life, promote sustainability, and drive economic development through the creation of new industries and employment opportunities (Caragliu et al., 2011; Linde, Sjödin, Parida, & Wincent, 2021; Vanolo, 2014). However, it is important to recognize that this transformation is not solely centered on technology. It encompasses a comprehensive understanding of the needs and aspirations of the people who inhabit and work in these cities. Therefore, in this book, we will explore the internal and external aspects of the smart city transformation, including data privacy and security concerns, citizen engagement and participation, and the importance of collaboration among different sectors and stakeholders. By considering these factors, this book aims to provide a comprehensive perspective on the transformation from a traditional city to a smart city, providing internal (organizational) and external (strategic) perspectives.

Internally, the transformation to a smart city involves changes in the way cities are planned, managed, and operated and the digital implementation act as a drive (and objective) in this transformation (Pittaway & Montazemi, 2020). Specifically, in the smart city trajectories, there is a greater emphasis on using data to inform decision-making processes. For example, smart city technologies can be used to monitor traffic flow, energy consumption, and air quality, among other things (Benevolo, Dameri, & Auria, 2016). These data can then be analyzed to identify areas for improvement and optimize urban services. Smart city technologies can also be used to improve communication and collaboration among different departments and stakeholders. By using digital platforms and tools, cities can streamline internal processes and reduce bureaucracy, making them more agile and responsive (Linde et al., 2021).

The transformation to a smart city also has significant strategic implications. Smart cities are designed to be more efficient, competitive, and attractive to businesses, investors, and visitors (Taylor Buck & While, 2017). By using digital technologies and data, smart cities can offer a range of benefits to their users, including faster and more reliable services, more sustainable and livable environments, and greater accessibility and inclusivity. These benefits can help cities attract and retain talent, businesses, and investment, enhancing their economic and social development (Christofi, Iaia, Marchesani, & Masciarelli, 2021; Kummitha & Crutzen, 2019; Lee, Florida, & Acs, 2004; Marchesani, Masciarelli, & Doan, 2022).

By considering both organizational and strategic perspectives in smart city development, this book will also explore how smart cities can enhance their competitiveness by improving their urban and economic environments, creating a conducive atmosphere for citizens, students, tourists, businesses, and entrepreneurs to thrive. In this regard, on the one side, smart city initiatives have the potential to foster innovation and entrepreneurship, generating new opportunities for startups and small businesses (Hollands, 2015; Sofronijevic, Milicevic, & Ilic, 2014). On the other hand, they also attract residents and tourists who are drawn to a dynamic environment and the multitude of opportunities it offers (Del Chiappa et al., 2016; Marchesani, 2022; Marchesani, Masciarelli, & Bikfalvi, 2023).

In doing so, smart cities can enhance their attractiveness to visitors by offering a wide range of digital services and experiences. Through the utilization of digital technologies and data, smart cities can provide personalized and interactive experiences to visitors, resulting in increased satisfaction and loyalty (Marchesani, 2022). For example, smart city technologies can enable the provision of real-time information on local attractions, events, and amenities, making it easier for visitors to navigate the city and discover new things (Del Chiappa et al., 2016).

In conclusion, this book will delve into these specific aspects of smart city development analyzing and discussing how the integration of digital technologies and data can not only improve efficiency and sustainability but also foster economic growth, innovation, and entrepreneurship. Additionally, we will explore how smart city initiatives can enhance the visitor experience, making cities more appealing and engaging for tourists. By examining these dimensions, our aim is to provide a comprehensive understanding of the potential benefits and opportunities that smart city transformations can offer.

1.2 Definition of Smart City

The term smart city has become increasingly popular in recent years, with cities around the world investing in digital technologies and data to improve their urban environments. However, despite the widespread use of the term, there is no widely agreed-upon definition of what a smart city actually is (Albino et al., 2015). This is somewhat of a paradox, as it is difficult to measure the impact and outcomes of something that is not well-defined.

The lack of a clear definition of a smart city is a significant challenge for policymakers, planners, and researchers. Without a clear understanding of what a smart city is, it is difficult to assess the effectiveness and impact of smart city initiatives. It also makes it challenging to compare and learn from different smart city projects and to identify best practices (Boes, Buhalis, & Inversini, 2015; Camboim, Zawislak, & Pufal, 2019; Linde et al., 2021; Pittaway & Montazemi, 2020).

In the literature, there are many different definitions of what a smart city is. However, most definitions share some common features, including the role of technology, efficiency, and sustainability. Here are some of the most prominent definitions:

- Caragliu et al. (2011, p. 50): "A city is smart when investments in human and social capital and traditional (transport) and modern (ICT) communication infrastructure fuel sustainable economic growth and a high quality of life, with a wise management of natural resources, through participatory governance."
- Bakici, Almirall, & Wareham (2013, p. 139): "Smart city as a high-tech intensive and advanced city that connects people, information, and city elements using new technologies in order to create a sustainable, greener city, competitive and innovative commerce, and an increased life quality."
- Nam and Pardo (2011b, p. 187): "A smart city infuses information into its physical infrastructure to improve conveniences, facilitate mobility, add efficiencies, conserve energy, improve the quality of air and water, identify problems and fix them quickly, recover rapidly from disasters, collect data to make better decisions, deploy resources effectively, and share data to enable collaboration across entities and domains."

These definitions highlight the importance of using technology and data to improve urban services and create more sustainable and livable environments. They also emphasize the importance of citizen participation and collaboration across different sectors and domains.

However, despite these commonalities, there are also significant differences between the proposed definitions. For example, some definitions focus more on the economic benefits of smart city initiatives, while others emphasize the environmental or social benefits. Some definitions also prioritize certain technologies or sectors over others.

This evolutionary nature in the definition of the concept of smart cities reflects the fact that they are complex and multifaceted phenomena that are constantly evolving. Despite their great potential, it is evident that there is still much to investigate and discuss regarding the role of smart cities in the cultural, social, and economic transformation of today's societies (Berrone, Ricart, & Carrasco, 2016; Christofi et al., 2021; Vanolo, 2014).

Throughout this book, we will strive to shed light on the internal processes and the impact of these smart city projects. By delving into the intricacies of these projects, this book aims to provide a comprehensive understanding of the dynamic and evolving nature of smart cities and their implications. Today, it is crucial to explore and analyze the cultural, social, and economic dimensions of smart city transformations to fully grasp their significance in contemporary societies.

1.3 Information and Communication Technologies (ICTs) and Smart City Trajectories

ICTs have played a fundamental role in shaping modern society, transforming the way we live, work, and communicate. In recent years, ICTs have become increasingly important in the context of cities, as urban areas around the world are adopting digital technologies to improve and guide governance and urban organization (Batty et al., 2012). The use of ICTs in cities has given rise to the concept of the smart city, a city that leverages digital technologies and data to create more efficient, sustainable, and livable urban environments (Lim, Edelenbos, & Gianoli, 2019; Maria Lluïsa Marsal-Llacuna, Colomer-Llinàs, & Meléndez-Frigola, 2015).

The integration of ICTs and other digital technologies into the urban environment is a key component of the smart city concept. Cities are using a range of ICTs, including sensors, data analytics, and mobile applications, to monitor and manage various urban systems, such as transportation, energy, and waste management. These technologies provide real-time data on urban conditions, enabling cities to make more informed decisions and optimize their services (Badii et al., 2017; Gagliardi et al., 2017; Thatcher, Burns, & Dalton, 2021).

The implementation of ICTs in cities is also having a significant impact on the lives of urban residents. ICTs are changing the way people interact with their cities, providing new opportunities for participation, engagement, and collaboration (El-Haddadeh, Weerakkody, Osmani, Thakker, & Kapoor, 2019; Gagliardi et al., 2017; Paskaleva, 2009). For example, digital platforms and tools enable citizens to provide feedback on urban services, report issues, and participate in decision-making processes (Vecchio, Secundo, Maruccia, & Passiante, 2019). ICTs are also providing new opportunities for social and economic development, creating new jobs and businesses in the digital economy (Kummitha, 2018).

Despite the potential benefits of ICTs in Smart Cities, their implementation is not without challenges. Cities face significant barriers in adopting and implementing new technologies, including financial constraints, technical challenges, and concerns around privacy and security. Additionally, the benefits of ICTs are not distributed evenly across urban populations, with some groups being left behind in the digital divide.

Despite these challenges, cities around the world are investing heavily in ICTs as a means of improving their urban environments. The use of ICTs in Smart

Cities is seen as a key driver of economic growth, social development, and environmental sustainability. Companies and governments are partnering to develop new technologies and services that can improve urban services and create new opportunities for citizens and businesses (Kummitha, 2018; Wiig, 2015, 2016).

The importance of ICTs in Smart Cities is also reflected in international city strategies. Many cities around the world are investing heavily in ICTs as a means of improving their competitiveness and attractiveness to businesses, users, and investors (Christofi et al., 2021). For example, the European Union promotes the guide Smart Cities' initiative to create a network of Smart Cities across Europe, using ICTs to improve energy efficiency, reduce carbon emissions, and enhance citizen participation.[1]

The implementation of ICTs in Smart Cities is not just about technology but also involves a fundamental shift in the way cities are managed and operated. Cities are adopting new governance models that emphasize collaboration, participation, and transparency (Meijer & Bolívar, 2016). By using digital platforms and tools, cities can engage citizens in decision-making processes, improve communication and collaboration among different departments and stakeholders, and increase transparency and accountability (Huertas, Moreno, & Pascual, 2021; Kumar, Mookerjee, & Shubham, 2018; Yuan et al., 2020).

The use of open data is also a key component of the smart city concept. Open data refers to data that is freely available to the public, without restrictions on its use or redistribution. Open data can play an important role in improving open governance, accountability, transparency, and citizen engagement (Gagliardi et al., 2017). By making data available to the public, cities can enable citizens, businesses, and researchers to develop new applications, services, and products that can improve urban services and create new opportunities.

In conclusion, the use of ICTs in Smart Cities is a key driver of economic growth, social development, and environmental sustainability. ICTs are transforming the way cities are managed and operated, enabling new forms of collaboration, participation, and transparency. However, the implementation of ICTs in Smart Cities is not without challenges, including financial constraints, technical challenges, and concerns around privacy and security. Despite these challenges, cities around the world are investing heavily in ICTs as a means of improving their urban environments and creating new opportunities for citizens and businesses.

1.4 Data Ecosystem and Key Actors

In modern cities, data has become an increasingly important resource for managing and optimizing urban systems. Cities around the world are leveraging data to improve their services, enhance citizen engagement, and promote economic

[1]EPRS: European Parliamentary Research Service. Belgium. Retrieved from https://policycommons.net/artifacts/1339578/mapping-smart-cities-in-the-eu/1949353/

growth. However, the management of data in cities is a complex and multifaceted issue that involves a range of key actors and stakeholders (Barham & Daim, 2020; Tiwana, 2014).

The management of data in cities involves the creation of a data ecosystem that includes a range of actors, such as policymakers, governance bodies, public managers, citizens, users, and providers (Ooms, Caniëls, Roijakkers, & Cobben, 2020). These actors work together to create, collect, process, and use data to improve urban services and create new opportunities for social and economic development.

Smart Cities initiatives aim to "provide more efficient services to citizens, to monitor and optimize existing infrastructure, to increase collaboration amongst different economic actors, and to encourage innovative business models in both private and public sectors" (Marsal-Llacuna et al., 2015, p. 618). To achieve these objectives, cities need to create a data ecosystem that is well-managed, secure, and transparent.

One of the key challenges in managing data in cities is ensuring that citizens and institutions collaborate effectively. Citizens are a critical part of the data ecosystem, as they generate a vast amount of data through their daily activities (Nesti & Graziano, 2020). However, citizens also have concerns about privacy and security and may be reluctant to share their data with institutions unless they are assured that their data will be protected and used responsibly.

To address these concerns, cities need to create a culture of trust and collaboration between citizens and institutions. This involves developing transparent and accountable data policies, engaging citizens in decision-making processes, and providing them with the tools and resources they need to participate effectively in the data ecosystem (Mouton & Burns, 2021; Nesti & Graziano, 2020).

In addition to citizen engagement, the management of data in cities also involves collaboration between different institutional actors. Policymakers, governance bodies, and public managers all play a critical role in the management of data in cities, as they are responsible for creating policies and strategies that guide the use of data in urban systems (Bakici et al., 2013; Yigitcanlar, Kankanamge, & Vella, 2020a). These actors need to work together to create a coherent and effective data ecosystem that can support the needs of citizens and businesses.

Finally, the management of data in cities also involves collaboration with data providers, such as telecommunications companies, software developers, and data analytics firms. These providers play a critical role in the creation and management of data in cities, as they provide the tools, infrastructure, and expertise needed to collect, process, and analyze data (Asthana & Dwivedi, 2020). Cities need to work closely with these providers to ensure that they have access to the data and tools they need to optimize their services and create new opportunities for social and economic development.

This preface shows that the management of data in cities is a complex and multifaceted issue that involves a range of key actors and stakeholders. Cities need to create a well-managed, secure, and transparent data ecosystem that can support the needs of citizens and businesses. This requires collaboration between citizens, institutions, and data providers to create a culture of trust and collaboration that can drive innovation and growth in modern cities. As cities continue to evolve and adapt to new technological and social changes, the management of data will become an increasingly important issue that will shape the future of urban development.

1.5 Smart City Dimensions

The concept of the smart city encompasses a range of different dimensions, each of which is critical to the success of the overall concept. The smart city dimensions are six conceptually distinct characteristics that cover all the major areas of the city, including the economy, mobility, governance, environment, living, and people (Albino et al., 2015; Vanolo, 2014). These dimensions are essential for creating a modern and sustainable urban environment that can meet the needs of citizens and businesses.

The first dimension of the smart city is the Smart Economy. This dimension is linked to a spirit of innovation, entrepreneurialism, the flexibility of the labor market, integration in the international market, and the ability to transform. The Smart Economy is essential for creating a dynamic and resilient urban economy that can adapt to changing market conditions and support the growth of businesses.

The second dimension is Smart Mobility, which is related to local and supra-local accessibility, availability of ICTs, and modern, sustainable, and safe transport systems. Smart Mobility is critical for ensuring that citizens have access to reliable and efficient transportation options, which can reduce congestion, pollution, and travel time.

The third dimension is Smart Governance, which is related to participation in decision-making processes, transparency of governance systems, availability of public services, and quality of political strategies. Smart Governance is essential for creating a democratic and participatory urban environment that can meet the needs of citizens and businesses.

The fourth dimension is Smart Environment, which is understood in terms of the attractiveness of natural conditions, lack of pollution, and sustainable management of resources. Smart environment is critical for creating a healthy and sustainable urban environment that can support the well-being of citizens and businesses.

The fifth dimension is Smart Living, which involves the quality of life, imagined and measured in terms of the availability of cultural and educational services, tourist attractions, social cohesion, a healthy environment, personal safety, and housing. Smart Living is essential for creating a livable and inclusive urban environment that can meet the needs of diverse communities.

The sixth and final dimension is Smart People, which is linked to the level of qualification of human and social capital, flexibility, creativity, tolerance, cosmopolitanism, and participation in public life. Smart People are critical for creating a diverse and innovative urban environment that can attract and retain the best talent and support the growth of businesses.

In the following chapters, each of these Smart City dimensions will be explored in more detail, with a focus on the specific strategies, policies, and technologies that can be employed to create a more sustainable, efficient, and livable urban environment. By addressing each of these dimensions in a holistic and integrated manner, cities can create a smart city that is responsive to the needs of citizens and businesses, and that can support long-term economic, social, and environmental sustainability.

1.6 Smart City Attractiveness

As anticipated, the concept of the smart city has become increasingly popular in recent years, as cities around the world are embracing new technologies, policies, and strategies to create more sustainable, efficient, and livable urban environments. However, the success of the smart city concept depends not only on its technological and policy dimensions but also on its attractiveness to various stakeholders, including businesses, citizens, tourists, students, and visitors (Crevoisier & Rime, 2021; Marchesani et al., 2022; Nakano & Washizu, 2021).

The smart city concept can influence the attractiveness of a city in several ways. First, the smart city can make a city more competitive and innovative, which can attract businesses and entrepreneurs. Smart Cities have a spirit of innovation, entrepreneurialism, and flexibility that can help businesses to thrive and grow (Kummitha, 2018). Additionally, Smart Cities often have modern and efficient infrastructure, which can help businesses to reduce costs and increase productivity (Odendaal, 2021).

Second, a smart city can make a city more livable and attractive to citizens. Smart Cities often have high-quality public services, efficient transportation systems, and a healthy and sustainable environment, which can improve the quality of life for residents (Trencher, 2019). Additionally, smart cities often have a vibrant cultural scene, with museums, theaters, and other cultural attractions that can make a city more attractive to residents.

Third, a smart city can make a city more attractive to tourists. Smart cities often have a range of tourist attractions, including cultural landmarks, natural attractions, and modern infrastructure, which can make a city more appealing to visitors (Buhalis, 2020; Del Chiappa et al., 2016). Additionally, Smart Cities often have efficient transportation systems and high-quality hospitality services, which can enhance the tourist experience.

Fourth, the Smart City can make a city more attractive to students. Smart Cities often have high-quality educational institutions, including universities and research centers that can attract students from around the world (Kummitha & Crutzen, 2019; Marchesani et al., 2022). Additionally, Smart Cities often have a vibrant and diverse cultural scene, which can provide students with unique opportunities for personal and professional development (Richard Florida, 2004).

To empower the attractiveness of a smart city, cities need to develop strategies and policies that address the needs and preferences of different stakeholders. This requires a holistic and integrated approach that considers the different dimensions of the smart city, including the economy, mobility, governance, environment, living, and people (Vanolo, 2014).

For example, to attract businesses, cities need to create a business-friendly environment that supports innovation and growth. This can involve the development of policies and strategies that encourage entrepreneurship, such as tax incentives, funding opportunities, and supportive regulatory frameworks (Muñoz & Cohen, 2016). Additionally, cities need to invest in modern and efficient infrastructure, including transportation systems, energy networks, and broadband connectivity, which can help businesses to reduce costs and increase productivity.

To attract citizens, cities need to create a high-quality living environment that supports the well-being of residents. This can involve the development of policies and strategies that ensure access to high-quality public services, such as health care, education, and housing. Additionally, cities need to invest in sustainable and healthy environments, including parks, green spaces, and efficient waste management systems, which can improve the quality of life for residents (Adler & Florida, 2021).

To attract tourists, cities need to create a unique and appealing tourist experience that showcases the city's cultural, natural, and historical attractions. This can involve the development of tourism infrastructure, such as museums, theaters, and cultural centers, as well as high-quality hospitality services, such as hotels, restaurants, and transportation systems (Del Chiappa et al., 2016; Marchesani, 2022).

To attract students, cities need to create a vibrant and diverse educational environment that supports learning and personal development. This can involve the development of high-quality educational institutions, including universities, research centers, and vocational schools. Additionally, cities need to invest in cultural and social infrastructure, including libraries, museums, and cultural events, which can provide students with unique opportunities for personal and professional growth (Richard Florida, 1995, 2004).

In conclusion, the success of the smart city concept depends not only on its technological and policy dimensions but also on its attractiveness to various stakeholders. To ensure the attractiveness of a smart city, cities need to develop strategies and policies that address the needs and preferences of different stakeholders, including businesses, citizens, tourists, and students. This requires a holistic and integrated approach that considers the different dimensions of the smart city, including the economy, mobility, governance, environment, living, and people (Caragliu & Del Bo, 2019; Christofi et al., 2021; Vanolo, 2014). By addressing these dimensions in a comprehensive and strategic manner, cities can create a Smart City that is attractive to a range of stakeholders and that can support long-term economic, social, and environmental sustainability.

1.7 The Competitive Edge of Smart City

The aforementioned discussions on smart cities and their characteristics, the role of technology and data within them, and the development of attractive dimensions and context for citizens, businesses, companies, tourism, and investors, open up a new vision of the city. No longer seen as a static entity anchored to its urban context but as a dynamic and competitive perspective, vying against other cities' strategies to ensure a competitive advantage and enhance its status in the global environment (Christofi et al., 2021).

Smart cities have a distinct competitive edge in comparison to traditional cities, owing to their ability to provide a wide array of services and instruments that effectively attract and retain businesses, citizens, and various stakeholders. These cities are recognized for their modern and efficient infrastructure, which facilitates the delivery of high-quality public services while nurturing a sustainable and health-conscious environment. Consequently, the residents' quality of life is significantly improved, fostering an environment conducive to business growth and development. Furthermore, smart cities exhibit a dynamic cultural landscape, foster an innovative environment, and uphold an evolving vision, rendering them increasingly appealing to both present and prospective residents and visitors alike.

The competition between Smart Cities has led to increased innovation and efficiency in urban services, as cities seek to differentiate themselves from their competitors. Cities are increasingly acting like businesses, with a focus on creating value for their stakeholders and delivering high-quality services at a competitive price.

This competition has led to the development of new technologies, policies, and strategies that can help cities to become more efficient and effective in delivering urban services. For example, cities are experimenting with new transportation systems, such as autonomous vehicles and smart mobility solutions, that can reduce congestion and improve mobility (Benevolo, Dameri, & D'Auria, 2016; Seyfi & Hall, 2020). Additionally, cities are investing in new technologies, such as smart grids and energy storage systems that can help to reduce energy consumption and improve the reliability of energy supply (Neirotti, De Marco, Cagliano, Mangano, & Scorrano, 2014).

The competition between Smart Cities has also led to increased collaboration and knowledge sharing between cities. Cities are increasingly working together to address common challenges, such as climate change, urbanization, and economic development. This collaboration has led to the development of new best practices, policies, and strategies that can help cities to become more sustainable, resilient, and competitive (Sorenson, Rivkin, & Fleming, 2006).

To maintain their competitive edge, Smart Cities need to continue to innovate and adapt to changing market conditions and stakeholder needs. This requires a focus on creating value for stakeholders, delivering high-quality services, and promoting innovation and efficiency in urban systems (Berrone et al., 2016; Hollands, 2015; Taylor Buck & While, 2017).

Today, the users of cities are regarded as a pivotal factor in the contemporary landscape of competitive cities. Smart cities, by orchestrating urban dynamics around their users, strive to establish an evolutionary advantage over less developed and less competitive cities. As previously discussed, smart cities offer a range of benefits that can attract and retain businesses, citizens, and other stakeholders. This competitive environment has fostered innovation and efficiency in urban services, leading to the development of new technologies, policies, and strategies that can help cities become more efficient and effective in delivering urban services (Caragliu et al., 2011; Paskaleva, 2009; Taylor Buck & While, 2017).

As we move forward, it's important to recognize that we are both the product and the user of this changing landscape. The competition between cities is not just about attracting businesses and tourists but about creating an environment that supports the growth and well-being of its residents.

Chapters 6 and 7 of this book will explore this perspective in more detail, providing practical advice to the reader on how to navigate the era of Smart Cities. Specifically, in this book, we will discuss the challenges and opportunities of living in a competitive city and provide insights into how to leverage the benefits of Smart Cities to improve the quality of life for residents and users as well as promote economic prosperity.

Part One
Internal Implementation of Smart City Ecosystems

Chapter 2

Digital Implementation in the Smart City Ecosystem

Abstract

This chapter provides an integrated overview of digital implementation in smart city ecosystems, focusing on the data ecosystem, the data value chain, and the digital implementation in smart cities. It acknowledges the opportunities and challenges presented by the growth of big data and its impact on delivering public services and coordinating local ecosystems in smart cities. This chapter introduces the concept of a smart city data ecosystem and explores the implementation of digital solutions, highlighting the role of local management and internal and external providers. It examines the digitalization of processes and services, technological infrastructure, and social media in the context of the smart city, emphasizing the interaction between local and national data ecosystems. This chapter incorporates interviews with smart city managers, policymakers, and technicians involved in the digital development process, offering valuable insights into the practical implementation and usage of data management within smart cities and the potential impact on urban development.

Keywords: Digital implementation; smart city; data ecosystem; data value chain; big data; public services; technological infrastructure

2.1 Introduction to the Smart City Data Ecosystem

In the digital era, innovation is a central concept that no longer exists solely at the micro level within the four walls of a company but expands at the macro level and across a multitude of partners and stakeholders connected to each other, contributing to the advancement of innovation ecosystems (Kummitha, 2018; Linde, Sjödin, Parida, & Wincent, 2021). To better understand the role of innovation ecosystems in the context of smart cities, it is essential to define and explain the concept of ecosystems and innovation ecosystems.

An ecosystem, in general terms, refers to a complex network of interconnected elements that coexist and interact with each other. In business and organizational

The Global Smart City, 15–30

Published under exclusive licence by Emerald Publishing Limited
doi:10.1108/978-1-83797-575-420231002

contexts, an ecosystem represents a network of organizations, individuals, and resources that interact and collaborate to create value through innovation, products, or services (Kapoor, 2018). An innovation ecosystem, on the other hand, is a more specialized system where various actors, such as businesses, government entities, research institutions, and individuals, collaborate and share knowledge to foster innovation and growth in a specific industry or domain (Adner & Kapoor, 2010).

Innovation ecosystems have embraced businesses and innovative sectors in recent years and have been at the center of academic and policy debates over the past decade. These systems have been particularly valuable for enterprises and organizations as they facilitate the exchange of information, resources, and ideas, leading to the development of new products, services, and business models. The open nature of innovation ecosystems allows for the inclusion of diverse actors, which contributes to increased creativity and problem-solving (Carvalho, 2015; Linde et al., 2021).

This vision, as anticipated, is shifting toward cities, which are adopting dynamics and strategies typical of businesses and, in turn, providing ecosystem opportunities for business-model innovation by bringing together multiple diverse actors (Fernandez-Anez, Fernández-Güell, & Giffinger, 2018). For instance, smart city ecosystems offer opportunities for collaboration among energy and electricity providers, municipalities, construction companies, and citizens in efforts to improve efficiency through novel multi-actor value propositions (Chauhan, Jakhar, & Chauhan, 2021).

Although the concept of innovation ecosystems has been widely studied in the context of businesses and organizations, its application to the realm of smart cities is still an emerging area of research. In particular, the current understanding of how ecosystem leaders orchestrate extended ecosystems to profit in dynamic and uncertain environments is not well-developed (Linde et al., 2021). This knowledge gap has significant implications for the decision-making processes and internal roles within smart city ecosystems.

Since smart cities continue to evolve, it is crucial to explore and understand the dynamics of the smart city data ecosystem. This ecosystem encompasses various stakeholders, including public and private entities, as well as citizens, who generate, collect, process, and utilize vast amounts of data to improve urban living. The smart city data ecosystem is characterized by a high degree of complexity and interdependence, as well as the need for collaboration and information sharing among diverse actors (Kummitha, 2018; Trencher, 2019).

In this chapter, we will seek to shed light on smart city data ecosystems and explain, with the help of policymakers and public managers, the role of data management, organization, and interpretation considering the data value chain in contemporary cities. In this chapter, we will also discuss the role of data ecosystem management in cities, as well as the various levels of digital implementation in cities that embrace different outcomes and models. In doing so, the data value chain model is useful to understand the complete cycle of data management, from collection to storage, processing, distribution, and analysis of data. This model helps to understand the value of data and to identify areas where improvements can be made. To ensure effective data ecosystem management in smart cities, it is necessary to adopt a holistic approach that considers all steps of the data value chain. Furthermore, it is important for smart cities to focus on collaboration and information sharing among stakeholders. This can include creating shared data platforms and promoting partnerships between public and private entities.

To better understand how data ecosystem management can influence the success of smart cities, this chapter also includes interview with policymakers and managers involved in smart city projects which emphasizes the essential role of data governance and a strategic approach to data management.

Thus, smart city data ecosystems and data management are essential to the success of smart cities. Throughout the rest of this chapter, the book will delve into each of these points in more detail, providing a comprehensive overview of the role of data ecosystems and data management in the success of smart cities.

2.2 Implementing the Data Value Chain

In today's information-driven world, data have become an essential asset for individuals, organizations, and governments. As cities become increasingly interconnected and intelligent through the use of advanced technologies, the importance of data management and utilization cannot be overstated. In this context, understanding and implementing the data value chain in smart cities is crucial for their sustainable development and efficient functioning (Valdez, Cook, & Potter, 2018).

The data value chain is a theoretical framework that describes the process through which raw data are transformed into valuable information and knowledge. Specifically, the data value chain refers to the process of data evolution, starting from its collection, then analysis, dissemination, and ultimately, its impact on decision-making (Curry, 2016; Löfgren & Webster, 2020; De Simone, Ceci, & Alaimo, 2022). By analyzing this chain, a decision support tool can be developed to model the sequence of activities that an organization undertakes to deliver a valuable product or service to the market. The value chain categorizes the general value-adding activities performed by an organization, enabling them to be comprehended and optimized. It consists of a series of subsystems, each with inputs, transformation processes, and outputs.

In the context of smart cities, the data value chain plays a vital role in optimizing urban management, enhancing service delivery, and improving the overall quality of life for citizens.

Smart cities are urban environments that leverage digital technologies, data, and advanced analytics to improve the efficiency of public services, promote sustainable development, and enhance the well-being of their inhabitants (Albino, Berardi, & Dangelico, 2015; Vanolo, 2014). As such, the effective utilization and management of data are pivotal to the successful implementation of smart city initiatives. Understanding and implementing the data value chain (see Fig. 1) is crucial for the success of smart city initiatives. This is because the value generated

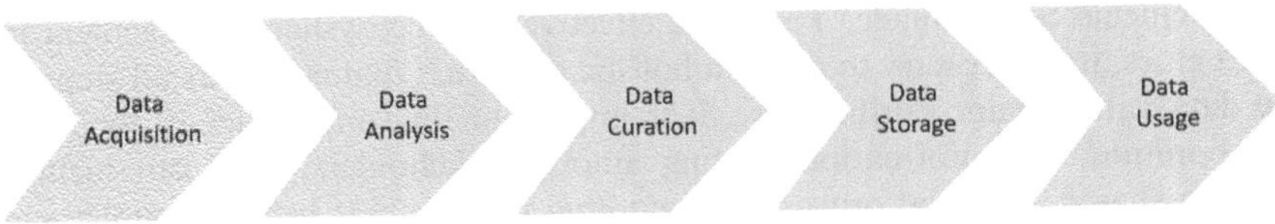

Fig. 1. Data Value Chain. *Source*: Created by author.

from the data can lead to innovative solutions and informed decision-making but also allows us to understand internally how data management is developed and who the main actors are in the processes of a smart city (Oztemel & Gursev, 2020; Pereira, Parycek, Falco, & Kleinhans, 2018).

Data acquisition is the first step in the data value chain, where raw data are created through various sources such as sensors, devices, social media, and other digital platforms and are gathered, filtered, and cleaned to generate an element that can be inserted in the Data Warehouse to carry out the necessary analysis. In smart cities, data acquisition occurs continuously as a result of the proliferation of connected devices and sensors embedded in urban infrastructure like traffic lights, public transportation systems, and utility networks.

These data sources generate massive amounts of information, often referred to as "big data," which can be harnessed to inform decision-making, optimize operations, and improve the overall quality of life in the city (Löfgren & Webster, 2020). The acquisition of these data is a direct consequence of the ongoing changes in cities and is an important key in the EU funding, as all outcomes and data related to smart city projects must be reported. For example, it emerged in one of the interviews that:

> The acquisition of data in smart city projects is important on three levels. Firstly, it is crucial when applying for European funding, as feasibility analyses and data are needed to justify the actual benefits and usefulness of these projects. Secondly, during the project development phase from acceptance to implementation, specific parameters (sustainable practices, digital accounting, and online reporting) must be met by companies collaborating with the city to participate in the bidding process. Finally, the services, infrastructure, and digital advances implemented in the smart city projects aim to generate a significant amount of data that can be used both by the European Union for project evaluation (with a guaranteed 5-year activity), and by the city for organizational and strategic purposes. This is why it is essential to organize and manage a vast amount of data throughout the smart city process.

Data analysis considers the modeling of data with the aim of making data useful for strategic purpose in the stage where data are examined, interpreted, and transformed into clear insight. In smart cities, data analysis techniques include descriptive, predictive, and prescriptive analytics, which allow stakeholders to understand historical trends, forecast future scenarios, and identify optimal strategies for decision-making.

Specifically, data analysis plays a crucial role in the data value chain, as it enables organizations to extract factual data and insights from the data they collect. This involves the use of various statistical techniques and algorithms to identify patterns, relationships, and trends within the data. Through this process, organizations can gain a deeper understanding of their operations, users, and the urban environment, and use this information to inform decision-making processes. Effective data analysis requires a combination of domain-specific knowledge, technical skills, and analytical thinking. It also involves the use of appropriate tools and technologies to manage and process large volumes of data efficiently. This allows smart cities to uncover hidden patterns, detect anomalies, and generate insights that drive innovation and improvement across various sectors such as transportation, public safety, and environmental management.

Data curation is a critical component of the data value chain in smart cities. Once data is collected from various sources, it must be processed to make it suitable for analysis. Data processing involves cleaning, organizing, and transforming raw data into a structured format that can be easily analyzed. In the context of smart cities, this process is particularly complex due to the large amounts of data generated from multiple sources but also the core competence embedded in the governance and public management. In line with this perspective, it is recognized that the full potential of the data available has not yet been realized due to the lack of internal competencies in data management within municipalities, as suggested in the following interview.

> For example, today we are able to receive a great deal of information from various internal sources such as intelligent traffic lights, public infrastructures, and sensors implemented in the city, regarding mobility. However, most of this information is used exclusively to monitor development in various areas of the city with specific offices and functions in collaboration with other designated offices (municipal police or PNRR reporting). We expect in the future, with digitally skilled human resources in the various offices, to expand the use and classification of this data in a strategic perspective to advance the city development and increase the relationship between the city and its citizens.

It assumes that there is a growing interest and awareness among public administrations regarding the potential of data curation in smart city projects. It is widely recognized that effective data management can lead to more informed decision-making, increased efficiency, and, ultimately, improved city perceptions.

Data storage refers to the process by which data are correctly clustered and stored in a scalable way, enabling organizations and users to quickly and efficiently access the data they need. In the context of smart cities, this process is particularly complex due to the large amounts of data generated from various sources and the need for the city to use these data as strategic tools.

Moreover, it also raises concerns related to data privacy and control. Organizations must ensure that the data they collect are stored securely and protected from unauthorized access, while also ensuring that they do not give too much power to data storage providers. To do that, there are two approaches to data storage for smart cities: one is to keep data in-house, while the second is to rely on a semigovernmental national provider for the smart transition of cities. For example, the following interview extract clarifies the circumstances where data storage is internal to the city rather than externalized to an external provider:

> The decision on data management lies with cities, which can manage it internally by taking risks regarding management and cybersecurity, but also the opportunities to use data that are not in the hands of third parties and therefore not tied to an external provider who then owns the data. Obviously, this choice concerns the management of cities but also the size and ability of the city to store and manage a large amount of personal data.

This approach of internal management is driven by organizational and strategic motivations; however, cities of different sizes have different needs and goals, which will also modify the strategy for data management and storage.

The final stage of the data value chain is *data utilization*, where the insights and knowledge derived from data analysis are applied to create value-added products, services, or solutions. In smart cities, data utilization can take many forms, such as the development of data-driven applications, the optimization of urban infrastructure, or the creation of new business models and revenue streams. It represent the outcomes of the data value chain, as it is where the true value of data is realized, and it has the potential to drive significant improvements in the efficiency, sustainability, and livability of urban environments (Christofi, Iaia, Marchesani, & Masciarelli, 2021; Linde et al., 2021).

The strategic utilization of data in cities is becoming increasingly important, and this trend is set to continue in the coming years. The potential benefits of data-driven solutions in smart cities are vast, ranging from improving traffic flow to reducing energy consumption and enhancing public safety.

In the second part of the book, we will analyze the strategic development of smart city projects and the impact they have on different types of users, such as businesses, entrepreneurs, tourists, citizens, and students. The success of smart city projects depends on effective data utilization, which requires a combination of technical expertise, domain-specific knowledge, and strategic vision. By leveraging data in a strategic and responsible manner, cities can create a more livable, equitable, and prosperous future for their inhabitants.

In conclusion, the data value chain is a crucial component of smart city initiatives, as it provides a framework for understanding and implementing the various stages involved in turning raw data into valuable insights and applications contributing to the data management in smart city projects.

2.3 Data Management in the City

As a consequence of the large amount of data previously discussed and the different approach into the data value chain, data management is becoming an increasingly critical factor in their success.

Effective data management in smart cities involves a range of complex tasks, from managing the massive amounts of data generated by Internet of Things (IoT) devices, sensors, and other sources to ensuring that these data are secure, accessible, and actionable for relevant stakeholders. This requires a combination of technical competences, domain-specific knowledge, and strategic vision, as well as the use of appropriate tools and technologies to process and analyze data effectively (Adler & Florida, 2021; Linde et al., 2021).

These competencies can either be developed internally by cities themselves or be provided by external software and service providers. For example, in some cases, cities may have existing technical and domain-specific expertise within their own organizations, which can be leveraged to develop and implement data management strategies for smart city projects. This may involve hiring specialized personnel, developing in-house data management software and infrastructure, and investing in training and development programs to ensure that staff has the necessary skills and expertise to manage data effectively.

Alternatively, municipalities may choose to partner with external providers who offer specialized data management services, software, and expertise. These providers may supply a range of services, including cloud-based data management and analysis, data visualization tools, machine learning (ML) algorithms, and standardized digital services to help cities manage their data more effectively.

As anticipated in the previous interview, the decision to develop data management competencies internally or to rely on external providers will depend on a range of factors, including the size and complexity of the smart city project, the availability of in-house expertise, and the budget and resources available. Ultimately, the key to successful data management in smart cities is to ensure that the right mix of technical, domain-specific, and strategic competencies are in place to enable cities to manage, analyze, and utilize data effectively for the benefit of their citizens. Specifically, a correct and structured data management empowers cities to:

(1) Developing a clear data strategy that outlines the goals, objectives, and priorities of the smart city initiative, as well as the data requirements and capabilities needed to achieve them.
(2) Establishing data governance frameworks that define the roles, responsibilities, and processes for data collection, processing, analysis, and utilization, as well as address key issues such as data privacy, security, and quality.
(3) Building data literacy and skills among stakeholders, including city officials, private sector organizations, and citizens, to ensure that they can effectively engage with and use the data generated by smart city initiatives.
(4) Promoting a data-driven culture within the smart city ecosystem, where data are viewed as a valuable resource that can drive innovation, inform policy-making, and improve the overall quality of life for citizens.

Thus, the successful implementation of the data value chain and data management in smart cities requires the collaboration and coordination of multiple stakeholders across the urban ecosystem. By recognizing the strategic and managerial role of data in smart cities, city governments and other stakeholders can ensure that their initiatives are well-placed to leverage the full potential of data and technology to create more efficient, sustainable, and livable urban environments. Section 2.4 explores in more detail the implementation of digital technologies in smart cities, focusing on four key areas: digitalization, digital services, technological infrastructure, and social media.

2.4 Digital Implementation in Smart Cities

Digital implementation is becoming an increasingly critical factor in the success of smart cities. As we discussed earlier, the data value chain and data management play a crucial role in enabling cities to collect, process, store, analyze, and utilize data effectively. However, the true value of data is realized when it is applied to create value-added products, services, or solutions.

In the context of smart cities, data plays a strategic and managerial role, as it enables city governments and other stakeholders to understand the needs and preferences of citizens and to develop and implement digital solutions that meet those needs. By leveraging data effectively, cities can optimize their operations, improve services, and enhance the quality of life for citizens (Marchesani, Iaia, Masciarelli, & Christofi, 2022).

On the other hand, the development of digital technologies and services generates a vast amount of data that can be used to inform decision-making processes and create more efficient and effective urban environments. This creates a virtuous cycle in which data and digital technologies complement each other, generating a continuous flow of information and services that benefit both cities and citizens (Kummitha, 2018).

The importance of digital implementation in smart cities cannot be overstated. Digital technologies and services are transforming the way cities are organized, and how they interact with citizens and other stakeholders. By embracing digital transformation, cities can optimize their operations, reduce costs, and improve the delivery of services to citizens (Thatcher, Burns, & Dalton, 2021).

Digital implementation also allows cities to optimize resource allocation, reduce waste, and improve sustainability. For example, the use of sensors and other IoT devices can help cities to monitor and manage energy consumption, reduce traffic congestion, and improve air quality.

Moreover, digital implementation can enhance the attractiveness and competitiveness of cities, making them more appealing to residents, businesses, and tourists. For example, providing high-quality digital services, cities can improve their status and attract citizens, tourists, students, and businesses, creating new opportunities for economic growth and innovation as will be discusses in the second part of this book (Christofi et al., 2021).

This implementation encompasses four major areas that comprise the development of the smart city, intersecting with various dimensions and improving the relationship between cities and users. Specifically, these areas pertain to the *digitization* of municipalities, the provision of *online services*, the development of advanced *technological infrastructure*, and the use of *social media* for cities' communications.

2.4.1 Digitalization

In recent years, we have seen that municipalities around the world have been investing in digital technologies to improve the efficiency and effectiveness of their operations. This has involved the use of automation, data analytics, and other digital tools to streamline processes, reduce costs, and enhance the quality of services provided to citizens (Vanolo, 2014).

The pandemic has accelerated this trend, as municipalities have been forced to adapt quickly to new ways of working and delivering services. This has required a greater focus on digital technologies, such as internal software, online portals, automized services, and virtual meetings, to facilitate communication and collaboration among employees and citizens (Chauhan et al., 2021).

One of the key benefits of digitalization is that it can improve the efficiency and effectiveness of municipal services. For example, the use of online portals can help citizens to access information and services more easily, reducing the need for physical visits to municipal offices. This not only saves time and money for citizens but also reduces the workload for municipal employees, who can focus on more complex tasks that require human input.

In addition, digitalization can contribute to a more transparent and accountable municipal government, by providing citizens with access to information and data that they can use to hold their government accountable. This can help to build trust and confidence in local government, and foster greater civic engagement and participation (Pereira et al., 2018).

However, digitalization also presents challenges, particularly in terms of ensuring the security and privacy of citizens' data as well as creates a distance between citizens and institutions (Chatfield & Reddick, 2019). For example, as regards the distance between citizens and institutions, one of our interviewees reported:

> The digitization of processes and services is undoubtedly improving efficiency and work in public offices, both internally and in relation to citizens. For example, in offices that previously required direct contact between the institution and the citizen, such as for births, marriages, or legal separations, digitization is improving both our organization of work and citizens' access to these services without physically going to the office. However, not all citizens are receptive to change, and while services must be hybrid to allow access for everyone, there is a sense from within

> that the border between institutions and citizens is being lost. In important moments in a person's legal life, the citizen still wants to have an institutional figure representing the state, and while useful, digitization could contribute to increasing the distance between institutions and citizens.

Thus, the process of digitalization, coupled with the development of online services, has significantly contributed to improving the performance of municipalities and cities, both in internal management and in relation to citizens. In this regard, the development of digital services directly contributes to the digitalization of cities.

2.4.2 Digital Services

The digitalization of municipalities has brought about a significant change in the way services are delivered to citizens. One of the most notable developments in this area is the rise of digital services, which have revolutionized the way citizens interact with their local government, making it more convenient and accessible (Benevolo, Dameri, & Auria, 2016; Marchesani & Masciarelli, 2023).

One of the most popular digital services offered by municipalities is the ability to access information and conduct transactions online, such as registering for residency, changing addresses, and accessing vital records. This has made it easier and more convenient for citizens to complete these tasks, reducing the need for physical visits to municipal offices.

Another area where digital services have made a significant impact is in the provision of mobility services. Many municipalities now offer online platforms and mobile apps for citizens to access public transportation services, such as bus and train schedules, real-time traffic updates, and route planning. This has made it easier and more convenient for citizens to navigate their city, reducing congestion and improving overall mobility (Marchesani, 2022; Seyfi & Hall, 2020).

The benefits of digital services in municipalities are numerous. They improve the efficiency and effectiveness of municipal operations, reducing costs and improving service delivery. Digital services also improve the quality of life for citizens, making it easier and more convenient to access services and information (Jungwoo Lee & Lee, 2014).

It assumes that by timing and accessible digital services, municipalities have improved the efficiency and effectiveness of their operations while also improving the quality of life for citizens. The adoption of digital services by municipalities, regardless of their size, has been driven by the push from national and EU institutions to improve digital services across all municipalities, further incentivizing the development and implementation of digital services in all municipalities.

2.4.3 Technological Infrastructure

While the digitalization of municipalities and the development of digital services are crucial in the implementation of smart cities, the development of adequate technological infrastructure is also a significant component. This is because public infrastructure, such as transportation, schools, offices, and public places, must constantly provide data to manage and improve their performance. Additionally, embedded in the smart city transformation, these structures must also adapt to embrace the principles of sustainability and efficiency that are central to the smart city paradigm (Gil-Garcia, Zhang, & Purón-Cid, 2016; Khatoun & Zeadally, 2016).

The development of technological infrastructure is crucial to the success of smart cities, as it provides the backbone for the collection, storage, and analysis of data that is necessary to make informed decisions about resource allocation and service provision. This includes the development of technologies such as sensors, IoT devices, and data analytics tools, which allow for the collection and analysis of data in real-time (Goodspeed, 2015).

Moreover, the development of technological infrastructure must also be aligned with the principles of sustainability and efficiency. This means that infrastructure must be designed to minimize energy consumption and environmental impact, while also providing high-quality services to citizens (Barham & Daim, 2020).

For example, Nearly Zero-Emission Building (NZEB) projects have been implemented in Italy and throughout Europe in smart city projects. NZEBs are buildings that have a very high energy performance, while the nearly zero or very low amount of energy required should be covered to a very significant extent by energy from renewable sources, including energy from renewable sources produced on-site or nearby.[1]

In conclusion, the development of technological infrastructure is a critical component of the implementation of smart cities. It provides the backbone for the collection, storage, and analysis of data that is necessary to make informed decisions about resource allocation and service provision. Additionally, the development of technological infrastructure must be aligned with the principles of sustainability and efficiency, ensuring that infrastructure is designed to minimize energy consumption and environmental impact while providing high-quality services to citizens. In this digital implementation, infrastructure is a central and tangible part of smart city projects.

2.4.4 Social Media

Social media has been adopted over the last decades by cities but has become a part of the smart cities implementation solely in the last year,[1] considering social media platforms to communicate with citizens and engage with them on a more personal level. This transformation shapes the way cities provide services, gather feedback, and communicate with their citizens (Yigitcanlar, Kankanamge, & Vella, 2020).

[1]https://european-union.europa.eu/

One of the most significant benefits of social media in smart cities is the responsiveness it provides. Social media platforms allow cities to respond to citizen feedback, and share real-time information, providing more personalized and efficient service. Additionally, social media platforms provide a medium for citizens to engage with their local government and participate in decision-making processes (Criado, Sandoval-Almazan, & Gil-Garcia, 2013).

However, there is an ongoing debate about the role of social media in smart cities, particularly regarding its integration into city projects. While social media has the potential to improve communication and engagement with citizens, for a long period and until 2023, it is not integrated into city projects, leaving it up to policymakers and public managers to decide how to incorporate it.

Furthermore, the use of social media in smart cities raises concerns about privacy and security. Cities must ensure that they have appropriate policies and procedures in place to protect citizens' personal information and ensure that social media platforms are used responsibly.

Despite these challenges, social media has the potential to transform the way cities communicate with their citizens, providing a more responsive and personalized service. It also has the potential to foster greater civic engagement and participation, enabling citizens to have a more significant say in the decision-making processes that affect their lives (Mossberger, Wu, & Crawford, 2013).

In conclusion, social media has become a crucial component in the implementation of smart cities, transforming the way cities communicate with their citizens. The use of social media platforms allows cities to be more responsive to citizens' feedback and complaints, providing a more personalized and efficient service. However, the integration of social media into city projects remains a challenge, and cities must ensure that they have appropriate policies and procedures in place to protect citizens' privacy and ensure the responsible use of social media platforms.

2.5 Local and National Data Ecosystem

The local and national data ecosystem is a complex network of data flows that operate at different levels, but are interconnected and interdependent. At the local level, data are collected from various sources, including sensors and other IoT devices, and is used to inform decision-making and improve the quality of services provided to citizens (Linders, 2012; Trencher, 2019).

At the national level, data are collected from multiple sources, including government agencies, private companies, and citizens themselves, and is used to inform policymaking and guide the development of national-level projects and initiatives. The national data ecosystem provides a framework for collaboration and information-sharing among municipalities, enabling them to learn from one another and share best practices.

However, as previously discussed, the implementation of smart city projects requires more than just data collection and analysis. Specifically, it requires a fundamental shift in the way public management is organized and structured, with a greater focus on collaboration, innovation, and citizen engagement (Allam & Dhunny, 2019). In this regard, cities are modifying their internal organization to meet current needs in terms of resources and human capital. For example, it emerged in an interview that:

> The management of cities has undergone profound changes over the years. Initiating change in a static organizational context anchored to the same way of working that existed in the 1980s has required significant financial resources and important decisions. For example, it was necessary to create a new city organizational unit that deals with digital transition, such as the "digital transition sector," which did not exist until a few years ago and now guides change in municipalities and public administrations. This is obviously a difficult job that requires not only strong local managerial drive and involvement in the project but also a national collaboration with the government and external actors such as software and service providers (in the case of Italy, the Maggioli Group) who work in unison towards the ultimate objective. Without a national government drive, it would be very difficult to initiate these projects both in terms of skills and resources. However, the role of local management is crucial, and this makes the difference between the various municipalities that, essentially, have the same access to resources.

A receptive public management is essential in pushing the digital transformation process, modifying the entire urban ecosystem. Smart city projects require a collaborative approach that involves multiple stakeholders, including government agencies, private companies, and citizens themselves (Supangkat, Arman, Nugraha, & Fatimah, 2018). This approach enables cities to leverage the collective knowledge and expertise of these stakeholders, creating a more efficient and effective urban ecosystem.

At the same time, local public management must also be able to navigate the complexities of implementing smart city projects in an environment that is not always receptive to change. Moreover, the implementation of smart city projects requires a high degree of citizen engagement and participation. Citizens must be informed and involved in the decision-making process, ensuring that their needs and preferences are taken into account.

Thus, it assumes that the local and national data ecosystem is a complex network of data flows that operate at different levels but is interconnected and interdependent. The implementation of smart city projects requires more than just data collection and analysis. It requires a fundamental shift in the way public management is organized and structured, with a greater focus on collaboration, innovation, and citizen engagement. A receptive public management is essential in pushing the digital transformation process, modifying the entire urban ecosystem. The implementation of smart city projects requires a long-term vision and a willingness to invest in new technologies and infrastructure, as well as a high collaboration between local management.

2.6 Data Storage, Security, and Challenges

As we have seen in this chapter, the rise of smart cities has brought about a new era in urban development, where technology and data play a crucial role in improving efficiency, sustainability, and overall quality of life for citizens (Caragliu & Del Bo, 2021; Chatfield & Reddick, 2019; Marchesani, Masciarelli, & Bikfalvi, 2023; Vanolo, 2014). One of the most critical aspects of this digital transformation is the management of data storage, security, and the challenges that arise within the smart city ecosystem. This section delves into the various aspects of data storage, security measures, privacy, ethical considerations, data governance and compliance, as well as the challenges associated with these areas. We will also explore future trends and innovations in the field of data storage and security for smart cities.

An effective and reliable data storage system is vital for the success of a smart city. It serves as the foundation upon which various applications and services are built, enabling the collection, processing, and analysis of massive amounts of data generated by sensors, devices, and other sources. The choice of data storage technologies depends on the specific needs of a smart city, with several options available such as cloud storage, local data centers, and distributed storage solutions (Jindal, Kumar, & Singh, 2020; Kiss & Muha, 2018; Li & Liao, 2018).

Cloud storage, for instance, offers a scalable and cost-effective solution for managing data, allowing cities to store data remotely and access it on-demand. Local data centers, on the other hand, provide a more centralized approach to data storage, with the advantage of increased control over the data's physical location and security (Hashem et al., 2016; Rahman et al., 2019). Distributed storage solutions, which involve spreading data across multiple locations and devices, can enhance the overall resilience and reliability of the system.

This discourse is not comprehensive and must be adapted to the specific characteristics of cities, which, for organizational or strategic reasons, may choose to either rely on external providers or government support for data storage, or internally organize data storage accordingly. For instance, one of the interviews revealed that:

> As a municipality, we are strongly pushing for this transformation, and we have made a very specific choice regarding data storage, management, and handling. We have decided not to outsource and, therefore, not to rely on external providers or government entities. There are two reasons behind this decision. Firstly, we want to avoid any ties or constraints with external providers that could potentially limit our long-term strategic planning. Secondly, we intend to explore multiple avenues by utilizing the wealth of data from various domains such as mobility, tourism, sensors,

> infrastructure, and more, to make informed decisions based on factual data. Of course, we are uncertain about the extent to which this will be possible. I am aware that many municipalities have outsourced this practice to mitigate security issues and cyberattacks. However, this depends on the internal strategy and the size of the municipality. From a strategic standpoint, we are currently proceeding along this path, which has yielded excellent results thus far.

In addition to selecting the appropriate data storage technologies, it is crucial to implement data security measures that protect the integrity, confidentiality, and availability of data within the smart city. This includes access control mechanisms that ensure only authorized personnel can access sensitive information, as well as data encryption techniques that protect data from unauthorized access or tampering. Cybersecurity measures, such as intrusion detection systems, firewalls, and regular security audits, are also essential to defend against cyber threats and maintain compliance with privacy regulations (Chatfield & Reddick, 2019; Christofi et al., 2021; Kiss & Muha, 2018; Ng & Wakenshaw, 2017).

In this line, privacy and ethical considerations are of paramount importance in the smart city context, as the collection, processing, and use of personal data have significant implications for citizens' rights and well-being. To address these concerns, smart cities must adopt privacy-by-design principles that prioritize the protection of personal data throughout the entire data lifecycle, from collection to storage, analysis, and eventual deletion (Löfgren & Webster, 2020; Lytras & Visvizi, 2018; Sharifi, Khavarian-Garmsir, & Kummitha, 2021; Ullah, Qayyum, Thaheem, Al-Turjman, & Sepasgozar, 2021). Additionally, ethical guidelines should be established to ensure that data are used responsibly and transparently, with the primary aim of enhancing the quality of life for citizens and promoting sustainable urban development.

Data governance and compliance play a critical role in the management of data within the smart city ecosystem, as they provide a framework for defining responsibilities, processes, and policies related to data handling. Effective data governance ensures that data are accurate, consistent, and secure, while compliance with data protection regulations, such as the General Data Protection Regulation (GDPR), helps to safeguard citizens' privacy and maintain public trust in the smart city's operations (Yuan & Li, 2019).

The challenges that smart cities face in data storage and security are manifold. One of the most pressing issues is managing the sheer volume of data generated by a multitude of connected devices and sensors, which can strain existing infrastructure and pose difficulties in terms of data processing and analysis. Additionally, securing the vast networks of IoT devices present in smart cities is a complex task, as these devices often have limited security capabilities and are vulnerable to cyberattacks.

Responding to emerging threats is another challenge that smart cities must address, as new technologies and attack vectors continue to evolve, requiring constant vigilance and adaptation of security measures. To overcome these challenges, smart cities must invest in the development of robust data storage and security frameworks, as well as foster collaboration among stakeholders to share best practices, resources, and expertise (Jindal et al., 2020; Linde et al., 2021; Tiwana, 2014). Looking toward the future, innovations in artificial intelligence (AI), machine learning (ML), and advanced storage technologies are expected to revolutionize the field of data storage and security in smart cities. AI-driven security solutions, for example, can help detect and respond to cyber threats more effectively, while new storage technologies, such as DNA-based data storage, have the potential to accommodate the ever-growing volumes of data generated by smart city systems.

Thus, it assumes that data storage, security, and the associated challenges are integral components of the smart city ecosystem, with significant implications for the success and sustainability of these urban environments. By addressing these issues head-on and embracing future trends and innovations, smart cities can unlock the full potential of data-driven solutions and create a better future for their citizens.

Chapter 3

Orchestrating the Implementation of the Smart City

Abstract

This chapter examines the six smart city dimensions that serve as pillars in smart city projects. These dimensions are crucial in the development and evaluation of smart city initiatives, representing key areas for consideration. This chapter offers a detailed analysis of the smart city ecosystem, focusing on the governance, environment, people, living, mobility, and economy dimensions. It challenges the prevailing media portrayal of the smart city strategy and engages in the current academic debate surrounding these dimensions. This chapter defines, discusses, and explains each dimension, incorporating case studies from cities such as Copenhagen, San Francisco, Lisbon, and Barcelona. It also includes interviews and factual data to highlight the internal implementation and objectives of the smart city within each dimension. This chapter provides a comprehensive understanding of the smart city ecosystem, its implementation, and the potential benefits and challenges associated with each dimension.

Keywords: Smart city; dimensions; governance; environment; people; living; mobility; economy

3.1 Understanding Smart City Dimensions

In the previous chapter, we discussed the concept of smart cities, their definition, and the importance of embracing this approach for sustainable urban development. In this chapter, the aim is to delve deeper into the various dimensions of smart cities and how vital it is to orchestrate these dimensions harmoniously to touch upon all crucial aspects of life and urban dynamics.

As Dirks and Keeling (2009) emphasized, organic integration of a city's various systems (transportation, energy, education, health care, buildings, physical infrastructure, food, water, and public safety) is essential in creating a smart city. Moreover, Carvalho (2015) stresses that infusing intelligence into each

The Global Smart City, 31–54

Published under exclusive licence by Emerald Publishing Limited
doi:10.1108/978-1-83797-575-420231003

subsystem of a city is not sufficient to develop a smart city. Instead, it should be treated as an organic whole.

In light of the complexity of managing the smart city concept in a holistic way, various researchers have attempted to break down the concept into smaller, more manageable features and dimensions. Giffinger, Fertner, Kramar, and Meijers (2007) is one of the first scholars to identify the six main components of a smart city: smart economy, smart mobility, smart environment, smart people, smart living, and smart governance. These components were later associated with different aspects of urban life and have become essential to recognize that these dimensions are interrelated and should be balanced to achieve the best outcomes for a smart city (Albino, Berardi, & Dangelico, 2015).

The six dimensions of a smart city are not isolated entities. Instead, they are interconnected, with each dimension influencing and being influenced by the others (Linde, Sjödin, Parida, & Wincent, 2021; Vanolo, 2014). The key to a successful smart city lies in the ability to orchestrate these dimensions effectively to create a cohesive urban ecosystem that addresses the challenges and opportunities of the 21st century.

One of the critical aspects of this orchestration is the role of information and communication technologies (ICTs) as we anticipated in Chapter 2. Nam and Pardo (2011a) argue that a truly smart city emerges when investments in human and social capital, along with ICT infrastructures, fuel sustainable growth and enhance the quality of life. While ICT is a vital component of smart cities, it is not the only factor. It is essential to recognize the importance of human capital and the necessity for smart governance that fosters creativity, diversity, and inclusivity. Regarding this matter, this perspective clearly emerges in the current trajectories of smart city development. For example, in an interview with a public manager, it emerged that:

> Obviously, the technology we have at our disposal as a city and the ability to operate on multiple fronts make us in line with the digital standards proposed at the country and European levels. The main problem is the management of this technology, which is why we need internal resources and high-skilled human capital to implement these practices internally (with the help of external suppliers). In this regard, the example of our municipality is perfect because we have established a dedicated office to bring together internal expertise and technology. We can say that until three years ago, technological innovation did not have a clear definition within a particular sector in our offices, but rather was a service, with a less structured dimension and half of the personnel we have today. However, we have now established a dedicated department that operates in this direction, and this has also been the winning card that, in retrospect, has allowed us to overcome the Covid period unscathed and initiate this digital transition, contributing to all six dimensions of the smart city.

Another crucial aspect of harmonizing the smart city dimensions is taking a human-centric approach. As Holland (2008) suggests, smarter cities begin with human capital rather than relying solely on ICT. By focusing on education and leadership and fostering an environment where entrepreneurship is accessible to all citizens, smart cities can ensure that the "smart people" dimension is fully realized (Neirotti, De Marco, Cagliano, Mangano, & Scorrano, 2014). This includes addressing barriers related to language, culture, education, and disability and encouraging creativity, flexibility, open-mindedness, and participation in public life (S. Y. Lee, Florida, & Gates, 2010; Marchesani, Masciarelli, & Quang, 2022).

Building on this discussion, in the upcoming sections of this chapter, we will delve deeper into each of the six smart city dimensions (see Fig. 2).

By examining these dimensions individually, we can better understand their role and potential in smart city projects. Through practical examples and insights from interviews with experts, we will highlight the importance of balancing these dimensions for a truly intelligent, sustainable urban environment.

As we continue our exploration of the various dimensions of smart cities, it is crucial to keep in mind the importance of harmoniously orchestrating these

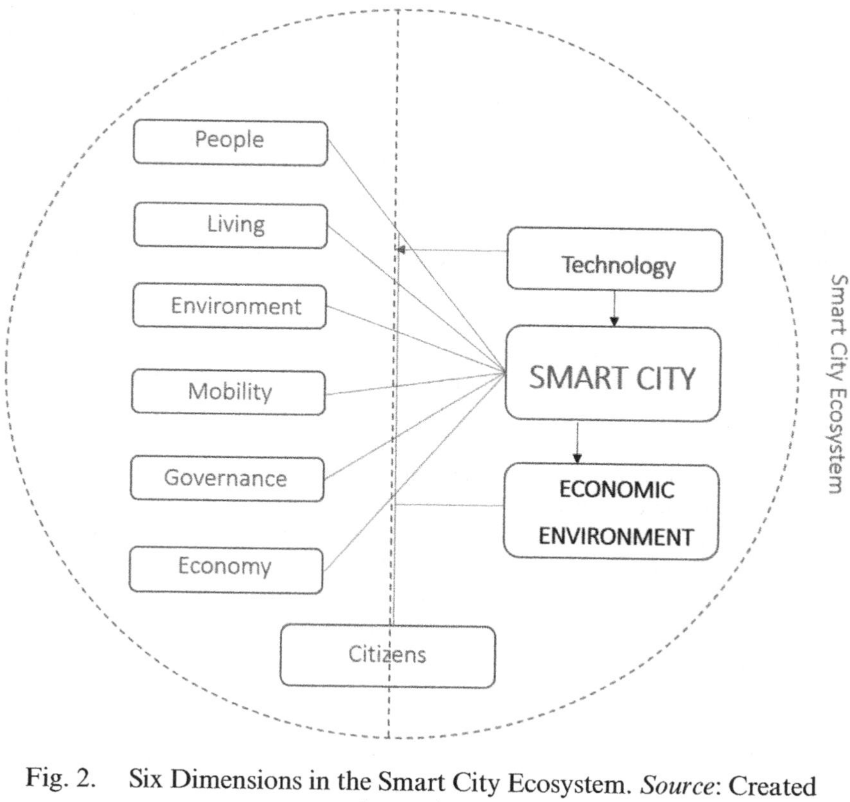

Fig. 2. Six Dimensions in the Smart City Ecosystem. *Source*: Created by author.

dimensions within the urban ecosystem. By doing so, we can create cities that not only leverage technology to improve efficiency and sustainability but also foster human capital, inclusivity, and creativity to enhance the overall quality of life for all residents.

3.1.1 Smart Governance

As previously discussed, we could say that the digital transformation is altering governance models in a disruptive way. Governance is the enabling environment that requires adequate legal frameworks and efficient processes to enable the responsiveness of government to the needs of citizens (UN Habitat, 2008). At the same time, it can also be defined as interaction and collaboration of different stakeholders in decision-making processes (Albino et al., 2015; Pittaway & Montazemi, 2020). The concept is commonly used to describe the action or manner of governing a state, an organization, or other constellation of actors. This shows that government and governance are related but different concepts.

Smart governance is defined as "*the capacity of employing intelligent and adaptive acts and activities of looking after and making decisions about the urban environment*" (Scholl & Alawadhi, 2016, p. 22), and it is "related to participation in decision-making processes, transparency of governance systems, availability of public services and quality of political strategies" (Vanolo, 2014, p. 887). According to these definitions, smart governance can be seen as the basis for smart, open, and participatory government. These concepts play a key role in the growing discourse on smart cities, so we may expect that ICTs play a key role in smart government as part of wider models of smart governance (S. Y. Tan & Taeihagh, 2020). And indeed, this concept emerges clearly also in interviews with smart city managers, who state that:

> In the development of smart city projects, governance and the way the city operates and interacts with its citizens is obviously fundamental component that ties different dimensions together and drives the development of smart cities. Without the strong push from the mayor and the administration, it would not have been possible to invest so many resources and efforts in smart practices in the city, which is now our strength in being one of the most digitally advanced cities in Italy. It is clear that these investments and efforts represent a great risk because the public may be more or less receptive, and we cannot be coercive in the use of digital services since the public is diverse and does not have homogeneous competencies. However, I can say that thanks to the vision of the current public administration and the creation of central offices and figures, the smart city project has had positive outcomes that are not only giving us great results but also driving us to push and invest even more in this direction.

Today, governments from all over the world are being challenged to become more innovative while at the same time reducing costs, operating in a connected environment, and engaging stakeholders in solving societal problems. In this context, the use of ICT in government has become a strategy for administrative reforms at all levels of government, and the pervasive use of ICTs in governance led to the creation of new governance perspectives such as e-government and digital government as a part of the smart governance dimension (Gagliardi et al., 2017; Pittaway & Montazemi, 2020; Sancino & Hudson, 2020). E-government or digital government can thus be seen as a transformational effort through the introduction of ICT in government organizations to achieve several objectives: improve the efficiency of public sector activities and service delivery but also transform government functions, internal organization, and practices through a citizen-centric approach; increase transparency; promote openness; and reduce corruption (Maria Lluïsa Marsal-Llacuna, Colomer-Llinàs, & Meléndez-Frigola, 2015; Maria Luisa Marsal-Llacuna & Segal, 2017; Neumann, Matt, Hitz-Gamper, Schmidthuber, & Stürmer, 2019; Shelton, Zook, & Wiig, 2015; Vu & Hartley, 2018). Therefore, governments around the world are adopting e-government strategies to improve the use of ICT. These emerging strategies mainly revolve around three added value "generators": administrative efficiency and interoperability, service improvement, and citizen centricity. In this sense, Tura and Ojanen (2022) argue that smart governance enables, with the help of ICT, the participation and collaboration of various actors in decision-making. For several authors (Abid, Marchesani, Ceci, & Masciarelli, 2023; Borrás & Edler, 2020; Gagliardi et al., 2017; Marchesani, Masciarelli, & Bikfalvi, 2023; Pinna, Masala, & Garau, 2017, 2022) this smart city governance makes use of new technologies and develops innovative governance combinations. Electronic platforms represent the use of ICTs to encourage citizen participation in decision-making, improving the provision of information and services, and enhancing transparency, accountability, and credibility (Mergel, Edelmann, & Haug, 2019). It is thus showing the possibility of a collaborative link between government and society, with ICT as an ally. After all, the main goal of smart city governance is not just the use of new technologies but the contribution of the urban environment that should focus on the community, network, and participants (Sancino & Hudson, 2020). In this sense, the pandemic has accelerated the relationship between governments, technology, and citizens, as reported in the interview:

> Technology is a useful means, and we have become fully aware of this during the pandemic. The only way to provide services to citizens was to become digital and push citizens to digitize not by choice but out of survival. In fact, many internal services that were already available, as well as new services to support citizens during the pandemic, could only be provided online. This has led many people who would never have approached digital technology before to transform their habits, and the city itself to set up services and plan projects tailored to citizens' needs.

It assumes that governance has been improved over time by receiving new features by coming up with smart governance which is the way government works with the participation of various stakeholders and the use of technology, bringing better citizen participation, and public and private partnerships. Accountability, cost reduction, linkages between the spheres of power, government-directed efforts, and innovation in public service provide higher quality public service delivery and consequent quality of life.

As a final outcome, smart governance aims to tie together all dimensions of the smart city and allows smart city projects to be undertaken. With smart governance, cities can improve transparency, reduce corruption, increase citizen participation, and promote partnerships between the public and private sectors. By leveraging ICT, smart governance helps enable smarter, more innovative, and more responsive cities.

3.1.2 Smart Environment

The concept of the smart environment is an essential component of smart city projects, as it encompasses the efficient management of energy and natural resources for achieving a better quality of life and a cleaner and more efficient environment. According to Vanolo (2014), the smart environment is understood in terms of the attractiveness of natural conditions, lack of pollution, and sustainable management of resources, suggesting that the smart environment is particularly related to the efficient management of energy and natural resources in the urban area.

Today, the concept of smart environment is increasingly gaining importance in the current society that is facing significant challenges such as overpopulation, lack of resources, and climate change. The increasing urbanization around the world is putting enormous pressure on natural resources, which is leading to environmental degradation, pollution, and depletion of natural resources (Chu, Cheng, & Yu, 2021). To counter these problems, smart cities are adopting practices that focus on the efficient management of energy and natural resources.

The smart environment dimension is critical in addressing these challenges because it promotes sustainable management of resources, attractive natural conditions, and reduced pollution levels. By adopting smart environment practices, cities can reduce their carbon footprint, improve their energy efficiency, and create a better quality of life for their citizens (Abid, Marchesani, Ceci, Masciarelli, & Ahmad, 2022; Colding & Barthel, 2017; Rohracher & Späth, 2014). These practices include the use of renewable energy sources, green buildings, efficient management of waste, and water resources, among others. For example, an important case already mentioned previously is that of the Nearly Zero-Emission Building (NZEB) concerning renewable energies and new infrastructures. In this regard, managers responsible for the allocation of funds have emphasized in interviews that:

> Currently, with the PNRR, the European Union requires us to make investments that bring a gain in terms of energy efficiency. It is important to note that this gain means not just improvement, as public works financed with smart city projects must bring an energy surplus. Therefore, these works should not only have zero impact but also generate additional clean energy that can be used by the city in its ordinary management. This approach is now a standard practice, and as an investment accountability office of the PNRR, we have the responsibility to continuously monitor and discuss these objectives, which, I repeat, are mandatory in the current city development plan through European funds for smart cities.

We assume that the efficient management of resources and the reduction of pollution levels will be central to the future development of cities, and that the direction taken by governments is to push and guide this transformation. This is because it is not only critical for the environment but also for the economic and social well-being of the citizens. The smart environment approach is, therefore, crucial in helping cities to achieve their sustainability goals and to address the challenges facing the world today, and the adoption of smart environmental practices is a significant step toward building sustainable and resilient cities that can withstand the challenges of the 21st century (Abid et al., 2022; Vanolo, 2014). For example, in Italy, environmental issues like air pollution, waste, and water management as well as climate change are rising in cities. The government has taken measures at the national level to mitigate these issues and reduce emissions but at the same time, environmental sustainability is also crucial part of the Sustainable Development Goals (SDGs) that Italy aims to achieve (Vanolo, 2014).

Although Italy's economy and society have developed unevenly across regions over the past decades, substantial differences remain in the progress of environmental and social development across cities (Leonelli, Marchesani, & Masciarelli, 2022). Thus, many Italian cities still face more traditional problems like lack of public services or deterioration of historical centers; they are redefining these issues in line with the smart city concept. Consequently, cities are reorganizing their agendas to transform into smart cities.

Cities shape the urban environment through infrastructures and buildings to enable future properties and meet the current migration from the urban areas to the cities. However, urban development consumes natural resources and energy, emits pollutants, and generates waste. Cities currently account for about 75% of the world's energy use and 70% of global CO_2 emissions, and these figures are rising as cities grow larger (Nasir, Duc Huynh, & Xuan Tram, 2019). The increasing intensity of urban metabolism and climate change are major sustainability challenges for cities, and the implementation of a smart environment dimension contributes to dealing with these issues.

Specifically, two main approaches to achieving a sustainable urban environment are considered. The first approach focuses on energy efficiency, renewable energy, green technologies, pollution control, and green management to reduce environmental impacts (Arabindoo, 2020; Gil-Garcia, Zhang, & Puron-Cid, 2016). The second approach centers on the efficient management of urban resources and utilities such as waste, water, transportation, and drainage to minimize pollution and improve resource quality (March & Ribera-Fumaz, 2016; Tachizawa, Alvarez-Gil, & Montes-Sancho, 2015). In this sense, smart mobility initiatives like reducing private vehicles and integrating transport also help decrease emissions and improve the environment as we can see in the next session.

Thus, it assumes that a smart environment is a key part of smart cities and aims to efficiently manage natural resources through technology and policies, and, in this sense, smart governance responsiveness to this challenge plays a central role in reacting to environmental contemporary challenges.

3.1.3 Smart Mobility

The need for improvements in urban infrastructure and subsystems has become a critical issue due to the rapid growth of urban populations and the high demand for quality of life. The increased demand for mobility has resulted in problems such as congestion and pollution (Cocchia, 2014). Following those issues, smart mobility has the potential to transform contemporary cities for the better as sustainable, integrated, and safe modes of transport must be considered in contemporary transport planning (Batty et al., 2012). Thus, smart mobility practices can offer users more transport options and more adaptable and affordable travel while reducing the reliance on private vehicles and promoting energy-efficient mobility (Benevolo, Dameri, & Auria, 2016).

In defining smart mobile, we must refer to a comprehensive concept that makes the transport network's sustainability more achievable due to the search for improvements in transport services, balancing the application of technology with social, economic, and environmental aspects (Vanolo, 2014). It is the ability to access transportation services from integrated platforms that aggregate the community and present intense processing of data from users to match the demand forecast.

Smart mobility is related to transport and the use of communication and information technologies to promote accessibility and increase the quality of life, and it includes the adoption of digital technologies that make mobility services in a city or territory more accessible and easier (Bakici, Almirall, & Wareham, 2013).

In implementing smart mobility practices, governments must be proactive and invest in sustainable transport solutions that promote energy saving. However, critical questions must be asked in terms of how the transition to smart mobility will be managed and how the benefits and any negative externalities of change will be governed (Docherty, Marsden, & Anable, 2018). The management of mobility by public and private agencies requires the adoption of intelligent and dynamic decision-making approaches capable of controlling various issues that change in real-time. Decision-making by responsible bodies must be linked to information such as the amount of data available, intelligent services, and energy results. They must also stick to issues that are specific to the reality of each city to implement relevant solutions that make mobility more intelligent (Vázquez, Lanero, Gutiérrez, & Sahelices, 2018). For example, a particularly noteworthy aspect of the interview discusses the opportunities and advantages of these practices.

> Our city policies are gradually adapting to the current mobility challenge and opportunities. The traditional means of transportation are changing, and cities are becoming smarter through the use of sensors, sustainable services, bike-sharing, and intelligent traffic lights, among others. For example, our city has a control-room that allows us to monitor the access to various public buildings, the movement of people on the streets, and in the city center, as well as in different areas. We work closely with the local police to ensure the safety of citizens and improve road safety.
>
> In addition to this, the control-room also helps us to monitor the vehicles passing through our city. We can check whether the vehicles are in compliance with the rules and regulations or not. We can also monitor the traffic flow and pollution levels (both noise pollution and air pollution) in real-time. This helps us to take necessary actions to control and reduce the pollution levels in our city.
>
> Moreover, the control-room helps us to monitor the speed of vehicles passing through the sensitive areas near schools. This allows us to take necessary actions to ensure the safety of children and improve road safety. By monitoring the traffic in real-time, we can offer better and more efficient services to the citizens of our city.

Thus, the urban mobility plan is a public management instrument that guides short-, medium-, and long-term projects, investments, and actions. It translates the objectives of improving local urban mobility into goals and strategic actions. Smart mobility requires a participatory process for preparing the plan, in which stakeholders identify challenges and propose improvement actions. They must also correlate the evolution in the usage dynamics of urban spaces with the transport system and consider the effect of new interventions on the natural and built environment and citizens' lives (Batty et al., 2012; Benevolo, Dameri, & D'Auria, 2016).

Despite internal organization and structural advancement, it is interesting to note that smart mobility is not only a means to solve internal problems, but it also offers a competitive advantage for the city when used strategically. Below is a practical example of this strategical perspective highlighted by a study conducted by Jan Henrik Nilsson (2019) on the city of Copenhagen:

Lessons from the City: Copenhagen, Denmark

The city of Copenhagen has leveraged the concept of smart mobility to enhance its transportation and attract new tourism, creating economic value for the entire area while simultaneously addressing mobility challenges. Specifically, cycling has been identified as a key aspect of smart mobility and urban planning efforts have been made to promote this mode of transportation. This includes the development of physical infrastructure such as bike lanes, commuting trails, intermodal solutions, bike-share schemes, and internet-based smart mobility.

It is worth noting that bicycle tourism cannot be viewed separately from everyday cycling practices carried out by the local population. While the regional bicycle culture is highly path-dependent, urban planning strategies and the development of bicycle infrastructure form the basic conditions for urban bicycle tourism to flourish. This opens up new opportunities for visitors' interactions with the city and the local community.

The local bicycle culture also creates awareness among tourists about possible cycling activities, which reinforces cycling as part of the destination identity. It becomes a unique aspect of the local experience that visitors are expected to participate in, contributing to a specific form of localness and generation an economic value for the city itself.

It assumes that smart mobility is an essential area of development for smart cities, as it impacts not only the mobility and internal management of a city but also its competitiveness and attractiveness. Smart mobility can be utilized as an economic and competitive engine for a city, generating value and contributing to its overall growth and development. Governments must be proactive in investing in sustainable transport solutions that promote energy-saving while ensuring that the transition to smart mobility is well-managed, and the benefits and any negative externalities of change are governed.

3.1.4 Smart Economy

The concept of a "smart economy" has emerged as a key aspect of the smart city paradigm. It embodies innovation, entrepreneurialism, labor market flexibility, integration into the international market, and the ability to transform business dynamics from a smart perspective. This section explores the smart economy dimension of smart cities, emphasizing its importance and elaborating on the factors that contribute to its development (Albino et al., 2015; Vanolo, 2014). It also explores how a well-developed smart economy creates an ecosystem that attracts and retains various stakeholders who interact with the city.

The smart economy is closely tied to a city's smart ecosystem and infrastructure, as well as the abundance of social and human capital. Together, these factors enable smart cities to develop more competitive business environments, attracting innovative companies, talent, students, knowledge, workforce, and investment (Vinod Kumar, 2020).

Consequently, smart cities contribute to establishing technology hubs that facilitate knowledge sharing through research centers, start-up incubators, accelerators, and innovation parks (Batty et al., 2012; Marchesani, Masciarelli, & Quang, 2022). Innovative cities and technology parks serve as natural magnets for establishing and promoting open innovation projects (Yun & Lee, 2019). For example, Morgan and Webb (2020) argue that when advanced IT infrastructures are developed locally through public–private partnerships, communities of lead users emerge in companies, public management, and university labs.

Moreover, the adoption of new technologies can rejuvenate markets and create economic benefits. By hosting firms that create such technologies, cities can gain economic superiority and competitiveness, attracting important company headquarters to the most developed cities (Camboim, Zawislak, & Pufal, 2019; Werner, 2002). This enhances the economic performance of cities both internally and externally when the technologies are effectively adopted and exported to other cities.

The prospect of shaping organizations according to European standards in terms of sustainability and innovation can potentially enhance their efficiency and help first to work in complex urban development. The ICT infrastructure of smart cities can also foster emerging and innovation-based business models, provided through regional, national, or multinational partnerships (Caragliu & Del Bo, 2019; Glaeser, 2000; Yigitcanlar, Kankanamge, & Vella, 2020).

However, public managers have highlighted the difficulty in matching public demand with private innovation, which brings both challenges and benefits to local firms. This difficulty is because the innovation brought by new technologies can be both challenging and beneficial, and local businesses may not always be able to respond promptly to these changes.

Despite the potential benefits of new technologies, dealing with contemporary technological innovations and meeting government requirements in terms of sustainability and digital reporting remains a challenge for local and national cities, as discussed in the previous section. This perspective is also evident in discussions with public managers who highlight the difficulty in creating a match between public demand and the level of innovation in the private sector:

> To date, as a city, we face significant challenges in engaging with local companies to implement these projects. There are very few companies that meet the European Union's requirements, and most of the time, we have to turn to foreign companies, mainly from the Netherlands, Poland, and Belgium. This is a paradox because there are many local companies that could collaborate and generate economic and social impact in the area. Unfortunately,

> their level of digitization holds them back, which also penalizes us as a city, resulting in longer lead times and a waste of resources.
>
> In the not-too-distant future, even local companies will have to adapt to this change because without these parameters, it will become increasingly difficult to participate in tenders and work with the public. In this case, the smart city promotes and encourages this transformation because if it doesn't, local companies will be severely penalized strategically. Therefore, they need to digitize and acquire the characteristics of sustainability and innovation, as in the rest of Europe.

Thus, despite the potential benefits of new technologies, dealing with contemporary technological innovations and meeting government requirements in terms of sustainability and digital reporting remains a challenge for local and national cities, as discussed in the previous section. However, the smart city is a driving force for this change, as it encourages local and national firms to adapt to these standards. Consequently, it is a challenge for local and national firms to shape their organizations according to European standards in terms of sustainability and innovation. This can potentially enhance their efficiency and help them work effectively in complex urban development.

In conclusion, it assumes that through the smart economy dimension, cities aim to become innovation-based, capital-based, and innovation hubs for businesses and human capital alike. Economic vibrancy embedded in the smart economy dimension is a key factor to affect a city's attractiveness and boosts the economic environment. When the smart economy dimension is well-developed, the city creates an ecosystem that draws in and retains all stakeholders who interact with it. By fostering innovation, collaboration, and the adoption of advanced technologies, smart cities can not only improve the well-being of their citizens but also create a sustainable, competitive, and thriving environment for businesses and talent to flourish.

3.1.5 Smart Living

Smart living is a crucial dimension in the development of smart cities. It encompasses initiatives that aim to improve health, education, social services, citizen participation, housing, and integration (Qonita & Giyarsih, 2022). However, the smart living dimension lacks clear conceptualization in the academic debate and presents challenges in terms of evaluating its impact on the territory (Vanolo, 2014).

The smart living dimension is not limited to the use of smart technology but emphasizes the quality of living produced by the independent implementation of smart technology under sustainable conditions. It aims to offer an urban context suited to the needs of its stakeholders by engaging with residents through advanced alternative or novel notions of urban and economic development, social inclusion, or greater urban planning (Camboim et al., 2019; Krishnan, Arumugam, & Maddulety, 2020; Richard Florida, 2002b). Citizens and users become the main actors in the smart transformation by becoming part of the remodeling and construction of the city itself. However, the relationship between citizens and the city is not always straightforward and presents social and environmental complexities. It is the city's responsibility to respond to and reduce this gap. One example that can relate to smart living practices is the city's support for its citizens during the pandemic, as reported in the following interview:

> During the pandemic, smart practices have become essential out of necessity. The city had to quickly understand the needs of its citizens and take drastic action to provide basic services for survival. This included not only education and healthcare services but also economic support to ensure that people could sustain themselves during difficult times. To address this challenge, we worked together with the government to provide meal vouchers and economic support in a fast and controlled manner that avoided people leaving their homes and creating crowds during the pandemic. We utilized the fiscal code card to distribute the support, and people in need adapted to this change, even those who were initially hesitant to use electronic money due to age or social construct.
>
> The city's response was prompt and effective, thanks to the use of current digital technologies. We were able to provide the necessary support to citizens and ensure their survival during the pandemic. ...It is clear that smart practices are crucial for responding to the needs of citizens in times of crisis, and we will continue to utilize innovative solutions to improve the quality of life for our citizens.

As cities continue to transform into smart cities, it is essential for the smart living dimension to adapt and follow suit. This is especially important as migration flows continue to move toward major cities, increasing the need for improved health services, education, security, and integration in the urban environment to enhance the quality of life (B. N. Silva, Khan, & Han, 2018a; Xie, Zhou, & Luo, 2016).

One of the most significant challenges that arise with an increase in population is the strain placed on healthcare systems. As the demand for healthcare services increases, hospitals and healthcare facilities need to adapt to meet the growing needs of the population (Feng & Yuan, 2023). For instance, cities like Singapore have implemented smart health-care solutions such as telemedicine, home monitoring, and mobile healthcare apps to provide convenient and accessible health-care services to their citizens (Sharifi, Khavarian-Garmsir, & Kummitha, 2021). These solutions not only improve the quality of health care but also reduce the burden on physical healthcare facilities.

Similarly, the education system must also adapt to accommodate the growing population. As cities become more crowded, schools must find ways to accommodate more students while maintaining a high standard of education. One solution is to implement digital technologies to enhance the learning experience such as virtual reality (VR), gamification, and online learning platforms. Cities like Helsinki, Finland, have embraced digital technology in their education systems, providing free access to online learning platforms and digital tools to all students (Hämäläinen, 2020).

Security and integration are also crucial factors that contribute to the livability of urban environments. As cities grow, so does the diversity of its population, making it essential to implement initiatives that promote social inclusion and integration. For example, the city of Amsterdam has implemented a program called "We Are Here" that focuses on providing shelter and support to undocumented migrants in the city, helping to integrate them into society and reduce social inequality (Hajer & Bröer, 2020).

In the current debate around smart cities, dimensions such as the environment, mobility, and governance clearly impact the current local and economic environment. However, the dimension of smart living is increasingly crucial for residents and the community as it includes a social and inclusive perspective of the city that directly impacts not only service provision and improvement but also living conditions (Adler & Florida, 2021; S. Y. Tan & Taeihagh, 2020).

Thus, it assumes that the smart living dimension plays an essential role in the balance of a city and is indispensable in supporting citizens. Its initiatives make a difference in the context of cities, and it is necessary to integrate them into urban planning and development strategies. Today, the smart living dimension must adapt and follow the transformation of cities into smart cities.

3.1.6 Smart People

In the smart city dimensions, the last (but not least) aspect that is often overlooked in the smart city discourse is the role of the smart people dimension. According to the definition presented by Vanolo (2014), smart people dimension is linked to the level of qualification of human and social capital, flexibility, creativity, tolerance, cosmopolitanism, and participation in public life. In other words, smart people are individuals who possess a wide range of skills and competencies, are open-minded, and actively engage in their community. The aim of contemporary cities is to embrace and promote these characteristics.

Smart cities can foster both human capital and social capital development. Since human capital considers the skills and competencies embedded in an individual or a group, smart city projects aim to attract and offer a social and inclusive environment tailored to the user. For example, the growth in a metropolitan area's concentration of college-educated residents is not only directly correlated with employment growth but also to other factors such as inclusiveness, services, nightlife, opportunity, culture, and lifestyle. Not surprisingly, the ability to meet people and make friends is one of the most important factors that determine our happiness in our lives and communities (Batabyal & Nijkamp, 2019; Richard Florida, 2002a, 2002b). Human beings crave interaction, but, as Putnam reminds us in the famous book *Bowling alone: The collapse and revival of*

American community, human beings have an innate desire for social interaction, but modern society often isolates us and makes it difficult to find satisfying social connections and support (Putnam, 2000). Smart cities can contribute to improving this aspect of contemporary urban and social life.

Thus, as anticipated, smart cities consist of various dimensions, such as smart mobility, smart environment, smart economy, and smart governance. However, smart people also play a central role interacting with each of these dimensions.

In fact, everything that relates to people's lives with passion, enthusiasm, and interest should be promoted by the city. Creating a lively and virtuous environment contributes to increasing the attractiveness of the city and the positive perception of its citizens and users (Richard Florida, 2012). A city that only focuses on the economy or governance risks losing touch with the public, which is both a user and a customer and must be attracted and pampered by the city.

Based on these perspectives, we can take the example of Florence and Barcelona. These two cities are known for their rich cultural heritage and innovative use of technology to promote it. Florence has numerous museums, galleries, and historic sites that attract millions of tourists every year. The city has also developed smart solutions to enhance the visitor experience, such as a mobile app that provides information and guidance on the city's cultural offerings (Nevola, Coles, & Mosconi, 2022). Similarly, Barcelona has a vibrant cultural scene that includes music festivals, art exhibitions, and street performances. The city has also implemented smart solutions to promote its cultural heritage, such as a digital platform that provides access to thousands of images and videos of the city's historic and artistic treasures (Bourliataux-Lajoinie, Dosquet, & del Olmo Arriaga, 2019).

London is another example of a city that has successfully attracted people and developed businesses by focusing on its attractions, innovation, and technology. The city is known for its iconic landmarks, such as Big Ben and the Tower of London, as well as for its innovative start-ups and tech companies combining and integrating those innovative advancements into smart city projects (Taylor Buck & While, 2017).

In conclusion, smart cities need to invest in the smart people dimension to create vibrant and attractive environments that benefit both citizens and users. By promoting a lively and virtuous environment that embraces people's passions and interests, cities can increase their attractiveness, foster innovation, and create new business opportunities. Therefore, it is important for cities to consider this dimension to drive human and social capital, attract and retain talented individuals, and provide them with the social and inclusive environment that they need.

3.2 Smart City Dimensions Empower City Development

In the previous chapter, we discussed the concept of smart cities and explored the various dimensions that constitute a smart city ecosystem. These dimensions encompass economic growth, mobility, sustainability, human capital, living standards, and effective governance. In this chapter, we delve deeper into the internal and organizational perspective of smart city development, focusing on the orchestration and balance of these dimensions. By doing so, cities can maximize the potential benefits and overcome challenges associated with implementing a holistic smart city approach.

In this section, we will analyze the various dimensions that empower city development within the smart city framework. Each dimension plays a crucial role in enhancing economic growth, improving mobility and connectivity, promoting sustainability, fostering human capital and well-being, and ensuring effective governance.

First, the smart economy dimension plays a pivotal role in empowering economic growth within a smart city. By harnessing technology and data-driven approaches, cities can create an environment conducive to innovation, entrepreneurship, and economic diversification. Smart economies facilitate the development of knowledge-based industries, attracting investments, fostering job creation, and enhancing productivity (Hollands, 2015; Kummitha, 2018; Marchesani et al., 2023). Through the integration of digital platforms, smart cities can stimulate collaboration among businesses, academia, and research institutions, promoting knowledge transfer and creating a thriving ecosystem for economic development (Langley & Leyshon, 2017; Sancino & Hudson, 2020). Second, the smart mobility dimension focuses on improving transportation systems, enhancing connectivity, and ensuring efficient movement within the city. By integrating smart technologies such as intelligent transportation systems, real-time data analytics, and smart infrastructure, cities can optimize traffic flow, reduce congestion, and enhance transportation efficiency. Smart mobility solutions, such as shared mobility services, electric vehicles, and smart parking systems, contribute to reduced carbon emissions, improved air quality, and increased accessibility (Benevolo, Dameri, & Auria, 2016; Benevolo, Dameri, & D'Auria, 2016). Additionally, smart mobility initiatives can enhance connectivity between different modes of transportation, promoting seamless and sustainable travel experiences for residents and visitors (Seyfi & Hall, 2020; Vázquez et al., 2018). Third, the smart environment dimension emphasizes the integration of sustainable practices and technologies to preserve and enhance the natural environment. Smart cities prioritize resource efficiency, waste management, and renewable energy solutions to minimize environmental impact. By leveraging technologies like smart grids, energy management systems, and sensor networks, cities can optimize energy consumption, reduce greenhouse gas emissions, and promote sustainable living. Furthermore, smart environmental initiatives involve water management strategies, green spaces, and urban planning approaches that prioritize ecological balance and resilience (Abid et al., 2022; Arabindoo, 2020; Linde et al., 2021). Fourth, the smart people and smart living dimensions focus on enhancing the quality of life, well-being, and social inclusivity within a smart city. Smart cities recognize the importance of human capital development, access to education, and the provision of social services. By leveraging digital technologies and data-driven approaches, cities can offer personalized and accessible services, such as e-learning platforms, health-care innovations, and citizen engagement platforms. Smart living initiatives also involve the creation of sustainable and inclusive communities, with a focus on affordable housing, public spaces, and cultural amenities that enhance residents' well-being and social cohesion (Aletà, Alonso, & Ruiz, 2017; S. Kumar, Mookerjee, & Shubham, 2018; Linde et al., 2021; Vázquez et al., 2018). Finally, the smart governance dimension emphasizes the adoption of data-driven decision-making processes, citizen participation, and transparent and efficient governance structures. Smart cities leverage technology to improve service delivery, enhance public safety, and streamline administrative processes (Meijer & Bolívar, 2016; Pittaway & Montazemi, 2020). By implementing smart governance practices, cities can leverage data analytics, artificial intelligence, and digital platforms to enable evidence-based policymaking, proactive urban management, and effective resource allocation. Smart governance also promotes transparency and accountability, fostering trust and engagement between the government and the citizens (De Guimarães, Severo, Felix Júnior, Da Costa, & Salmoria, 2020; Lnenicka et al., 2022).

By understanding the interconnections and interdependencies among these dimensions, cities can orchestrate and balance them effectively. The success of a smart city lies in the ability to harmonize economic growth with sustainable mobility, environmental stewardship, human well-being, and efficient governance. Achieving this balance requires integrated planning, collaborative partnerships, and data-driven approaches (Linde et al., 2021; Tiwana, 2014; Vanolo, 2014).

In the next chapter, titled "Orchestrating and Balancing Smart City Dimensions," we will delve deeper into the strategies and best practices for achieving this harmony. We will explore how cities can integrate different dimensions, leverage technology and data, and engage stakeholders to create a cohesive smart city ecosystem. Through practical examples and case studies, we will highlight successful models of dimension orchestration and provide insights for future smart city projects.

As cities continue to evolve and face new challenges, the orchestration and balance of smart city dimensions become increasingly critical. By embracing a holistic approach, cities can create intelligent, sustainable, and inclusive urban environments that enhance the overall quality of life for their residents. The next chapter aims to provide an overview for achieving this vision, setting the stage for the implementation of effective strategies and the realization of the full potential of smart cities.

3.3 Orchestrating and Balancing Smart City Dimensions

Within this chapter, we explored the various dimensions that constitute a smart city ecosystem and their significance in empowering city development. However, it is essential to recognize that these dimensions are not isolated entities but rather interconnected and interdependent. To achieve the full potential of a smart city, it is crucial to adopt a holistic approach that orchestrates and balances these dimensions effectively.

The interconnections between smart city dimensions create a synergistic effect, where advancements in one dimension can enhance and support progress in others. For example, the development of a smart economy can drive job creation and economic growth, which, in turn, can improve the quality of life for residents and promote sustainable living (Linde et al., 2021; Tiwana, 2014; Vanolo, 2014). Likewise, investing in smart mobility can enhance connectivity and accessibility, leading to increased economic opportunities and improved social inclusivity. To orchestrate and balance smart city dimensions, cities need to adopt strategic approaches that facilitate collaboration, integration, and optimization. Here are three key strategies for achieving effective orchestration and balance.

3.3.1 Integrated Planning and Policy Frameworks

To ensure effective orchestration and balance of smart city dimensions, cities should develop comprehensive and integrated planning frameworks that consider all dimensions simultaneously (Camboim et al., 2019; Gil-Garcia et al., 2016). This approach allows cities to align the objectives and actions of each dimension, creating a cohesive and synergistic approach to smart city development. By integrating smart city goals into urban planning processes, cities can optimize resource allocation and minimize redundancies. For example, when planning infrastructure projects, cities can consider how the smart mobility dimension can be incorporated to enhance transportation systems, while also considering the impact on the smart environment dimension by incorporating sustainable design principles. Integrated planning frameworks enable cities to identify potential synergies and trade-offs between dimensions, ensuring that actions taken in one dimension positively contribute to others. For instance, a policy aimed at promoting economic growth through innovation and entrepreneurship can also have positive effects on the smart people dimension by creating job opportunities and fostering human capital development (Bibri & Krogstie, 2020; Fortino, Russo, Savaglio, Shen, & Zhou, 2018; Matos, Vairinhos, Dameri, & Durst, 2017).

> **Lesson from the city:** San Francisco, United States
>
> An exemplary city that has successfully implemented integrated planning and policy frameworks to drive its smart city initiatives is San Francisco, located in the United States. The city's approach has been centered around key dimensions such as transportation, sustainability, and digital innovation, showcasing the benefits of a cohesive and synergistic strategy.
>
> (J. H. Lee, Hancock, & Hu, 2014; Yigitcanlar, Han, Kamruzzaman, Ioppolo, & Sabatini-Marques, 2019)
>
> In this line, in the realm of transportation, San Francisco has adopted forward-thinking policies to promote the use of electric vehicles (EVs) and enhance public transportation infrastructure. By establishing charging infrastructure and incentivizing EV

> adoption, the city has successfully reduced carbon emissions and encouraged sustainable transportation alternatives. Additionally, San Francisco has integrated smart technologies and data-driven solutions into its public transportation systems, offering real-time transit information, intelligent traffic management, and smart parking solutions to improve mobility and alleviate congestion.
>
> San Francisco's commitment to sustainability is evident through its policies and programs aimed at fostering green building development and utilizing renewable energy sources. By incorporating energy efficiency measures, water conservation initiatives, and effective waste management practices, the city has created a more sustainable urban environment. These efforts not only reduce the ecological footprint but also contribute to an improved quality of life for its residents.
>
> (J. H. Lee et al., 2014)
>
> Embracing digital innovation, San Francisco has harnessed the power of technology and data to enhance service delivery, optimize efficiency, and encourage innovation. The city has established open data platforms, employed smart city analytics, and implemented digital engagement platforms to promote citizen participation. By leveraging these digital tools and fostering innovation ecosystems, San Francisco has created an inclusive and connected city that empowers its residents and businesses.
>
> (Adler & Florida, 2021; J. H. Lee et al., 2014; Yigitcanlar et al., 2019)
>
> San Francisco's success in integrating planning and policy frameworks across multiple dimensions showcases the importance of aligning goals and actions. By considering the interdependencies between transportation, sustainability, and digital innovation, the city has optimized resource allocation and minimized redundancies. San Francisco's comprehensive planning framework serves as a practical example of how cities can orchestrate and balance smart city dimensions to create a sustainable and inclusive urban environment that leverages technology for the benefit of its residents.

3.3.2 Collaborative Partnerships and Stakeholder Engagement

Effective orchestration of smart city dimensions necessitates collaboration and engagement among various stakeholders. This includes government agencies, businesses, community organizations, academia, and citizens. Collaborative partnerships enable cities to leverage diverse expertise, resources, and perspectives to address complex challenges and develop innovative solutions. Engaging stakeholders in the decision-making process promotes inclusivity, ownership, and a shared vision for the city's future. By involving citizens, for example, in the cocreation of smart city initiatives, cities can ensure that solutions are tailored to the specific needs and aspirations of the community. Collaboration with businesses and academia can bring technical expertise and research capabilities, driving innovation and implementation of cutting-edge technologies (Abid et al., 2022; Lei, Ye, Wang, & Law, 2020; Zhao, Fashola, Olarewaju, & Onwumere, 2021).

Collaborative partnerships also facilitate knowledge sharing and learning among stakeholders. By exchanging best practices and lessons learned, cities can accelerate their smart city development journey and avoid potential pitfalls. Furthermore, partnerships with community organizations can ensure that the social dimensions of a smart city, such as equity and inclusivity, are adequately addressed in the planning and implementation processes.

> **Lesson from the city:** Lisbon, Portugal
>
> Lisbon has been proactive in fostering collaboration and engaging stakeholders in its smart city development. The city has established the Lisbon Urban Data Laboratory, which serves as a collaborative platform for various stakeholders, including local government, businesses, and academia. This platform enables these entities to work together in co-creating and implementing smart city solutions.[1]
>
> Through the Lisbon Urban Data Laboratory, stakeholders collaborate on projects focused on key smart city areas such as energy, waste management, mobility, and social inclusion. By involving diverse stakeholders, Lisbon ensures that multiple perspectives and expertise are considered, leading to more comprehensive and effective solutions. This collaborative approach also helps to leverage resources, share knowledge, and avoid duplication of efforts.
>
> Moreover, Lisbon has actively engaged its citizens in the smart city development process. The city has implemented the Smart Lisbon platform, an online platform that encourages citizen participation and contribution. Through this platform, residents can provide their ideas, suggestions, and feedback on various smart city initiatives. This engagement empowers citizens to play an active role in shaping the future of their city and ensures that solutions are tailored to meet their specific needs and aspirations.[1]
>
> (Albuquerque et al., 2021)

[1]EPRS: European Parliamentary Research Service. Belgium. Retrieved from https://policycommons.net/artifacts/1339578/mapping-smart-cities-in-the-eu/1949353/

> The collaborative partnerships and stakeholder engagement in Lisbon's smart city initiatives have yielded significant benefits. By involving businesses, academia, and citizens, the city taps into a wealth of expertise and resources, enabling the implementation of cutting-edge technologies and innovative solutions. The active engagement of citizens fosters a sense of ownership and inclusivity, ensuring that the smart city initiatives align with the priorities and values of the community. Lisbon's collaborative approach has resulted in tangible outcomes. The city has made remarkable progress in areas such as energy efficiency, mobility, and social inclusion. For example, initiatives like smart lighting systems, electric vehicle charging infrastructure, and community-driven programs have been successfully implemented. These efforts have not only improved the quality of life for residents but have also positioned Lisbon as a leading smart city on the global stage.
>
> (Camboim et al., 2019)

In this line, Lisbon's case exemplifies the importance of collaborative partnerships and stakeholder engagement in driving smart city initiatives. By involving diverse stakeholders and actively engaging citizens, Lisbon has been able to tap into collective knowledge, resources, and perspectives to develop innovative and sustainable solutions. The collaborative approach has not only led to successful implementation but has also fostered a sense of ownership and inclusivity, ensuring that the smart city transformation is driven by the needs and aspirations of the community.

3.3.3 Data Sharing and Interoperability

Data are a crucial resource for smart city operations and decision-making. Cities should prioritize data sharing and interoperability among different dimensions to enable comprehensive analysis and informed decision-making. Breaking down data silos and establishing open data platforms are essential steps in unlocking the full potential of data-driven approaches and facilitating cross-dimensional integration. By enabling data sharing, cities can combine data from various sources and dimensions to gain a holistic understanding of their urban challenges and opportunities. For instance, by integrating data from smart mobility systems with environmental sensors, cities can analyze the impact of transportation on air quality and identify strategies to reduce emissions (Dewi, Hidayanto, Purwandari, Kosandi, & Budi, 2018; Löfgren & Webster, 2020; Oztemel & Gursev, 2020; Pittaway & Montazemi, 2020). Interoperability is also critical in enabling the seamless exchange and integration of data between different systems and stakeholders. Standardized data formats, protocols, and interfaces allow for efficient data flow and analysis across dimensions. This interoperability empowers cities to generate valuable insights, improve efficiency, and enhance the overall performance of their smart city initiatives.

Lesson from the city: Barcelona, Spain

Barcelona, located in Spain, has emerged as a leading smart city by embracing data-driven approaches and prioritizing data sharing and interoperability among different dimensions of the city. The city has implemented the Sentilo platform, an open-source initiative that serves as a data-sharing hub for various smart city systems.[2]

One notable aspect of Barcelona's approach is its integration of data from sensors, devices, and applications across different dimensions, including transportation, energy, and environment. For example, the city has deployed smart sensors to collect real-time data on traffic patterns, air quality, and energy consumption. By combining data from these diverse sources, Barcelona gains a comprehensive understanding of urban challenges and opportunities. Through the Sentilo platform, Barcelona enables stakeholders to securely access and share data, fostering collaboration and innovation (Bakici et al., 2013; Bourliataux-Lajoinie et al., 2019). Government agencies, businesses, research institutions, and citizens can leverage the platform to develop data-driven solutions tailored to the city's needs. This collaborative approach ensures that data is harnessed effectively to address urban challenges and improve the quality of life for residents.

Barcelona's focus on data sharing and interoperability has facilitated cross-domain analysis, allowing the city to explore connections and synergies between different dimensions. For instance, by integrating data from transportation systems with environmental sensors, Barcelona can analyze the impact of mobility on air quality and develop strategies to reduce emissions. This interconnected approach helps optimize urban services, enhance sustainability, and create a more livable environment for its residents. Furthermore, Barcelona's emphasis on open-source initiatives like Sentilo fosters innovation and knowledge sharing (Sinaeepourfard et al., 2016). The platform enables developers and entrepreneurs to build applications and services based on shared data, driving economic growth and technological advancements. By encouraging a collaborative ecosystem, Barcelona leverages the expertise and creativity of a wide range of stakeholders to tackle urban challenges effectively (Bakici et al., 2013; Camboim et al., 2019; Nesti & Graziano, 2020; Sinaeepourfard et al., 2016).

The success of Barcelona's data sharing and interoperability initiatives can be attributed to the city's holistic approach to smart city development. By integrating data from various dimensions, engaging stakeholders, and fostering collaboration, Barcelona has

> created a thriving ecosystem that harnesses the power of data to transform urban life. Overall, Barcelona's case study exemplifies the importance of data sharing and interoperability in smart city initiatives. By breaking down data silos and establishing open platforms, cities can unlock the full potential of data-driven approaches, foster collaboration, and drive innovation for the benefit of their residents.

Drawing upon the aforementioned examples and in-depth discussions, it becomes evident that the orchestration and balance of smart city dimensions hold immense significance for cities aspiring to fully realize the potential of their smart city initiatives. By strategically integrating and aligning these dimensions, cities can foster a comprehensive and synergistic approach to urban development, thereby unlocking a myriad of opportunities and reaping substantial benefits.

Firstly, by orchestrating these dimensions, cities can leverage their unique characteristics and strengths to maximize the impact of their smart city initiatives. Every city has its distinct identity, economic landscape, natural resources, and societal makeup. By understanding and considering these intrinsic characteristics, cities can tailor their strategies and investments to align with their specific needs and goals (Coca-Stefaniak, 2021; Tiwana, 2014; Vanolo, 2014). For instance, a city known for its technological innovation can focus on developing a robust smart economy dimension, while a city with abundant green spaces can prioritize the smart environment dimension to enhance sustainability. Moreover, orchestrating smart city dimensions offers cities the opportunity to address interconnected challenges and seize synergies. For example, integrating the smart mobility dimension with the smart environment dimension can lead to solutions that reduce congestion, improve air quality, and promote sustainable transportation options. By recognizing and capitalizing on these interdependencies, cities can create comprehensive and integrated solutions that enhance the overall performance and resilience of their urban systems. Additionally, effective orchestration of smart city dimensions enables cities to optimize resource allocation and minimize redundancies. By taking a coordinated approach, cities can avoid fragmented and isolated initiatives, ensuring that investments and efforts are directed toward common objectives. This not only improves the efficiency and effectiveness of smart city projects but also maximizes the return on investment for both the city and its stakeholders. Finally, by successfully orchestrating these dimensions, cities can create a unique and attractive urban environment that fosters innovation, economic growth, and quality of life. A well-balanced and harmonious integration of smart city dimensions can enhance the city's competitiveness, attract talent, and stimulate economic activities. It can also improve the livability and well-being of residents by providing sustainable infrastructure, equitable access to services, and a vibrant community fabric.

In conclusion, orchestrating and balancing smart city dimensions is crucial for cities to unlock their full potential and capitalize on the opportunities presented by the smart city paradigm. By understanding their own characteristics and aligning their strategies with the dimensions that best represent and complement their city, urban centers can create a thriving ecosystem that benefits residents, businesses, and the overall urban environment. The effective orchestration of smart city dimensions will play a pivotal role in shaping the cities of the future, enabling sustainable, inclusive, and prosperous urban environments for generations to come.

Notes

1. https://urbandatalab.pt/index.php/the-lab/outputs/lisbon-urban-data-labs
2. *Sentilo* website: https://ajuntament.barcelona.cat/digital/en/digital-transformation/urban-technology/sentilo and https://connecta.bcn.cat/

Chapter 4

Geographic Patterns in Smart City Implementation

Abstract

This chapter investigates the geographical variations and approaches that shape the implementation of smart city projects on a global scale. Recognizing the significant influence of contextual factors on smart city initiatives, this chapter aims to comprehend the dynamics that drive the objectives and approaches of smart city projects across different countries. This chapter provides an overview of the strategic and geographical factors that impact smart city implementation, considering organizational drivers and practices, such as the Hofstede model in context. It explores the role of citizen-based, technology-based, and decision-making-based development in smart city implementation. Moreover, this chapter examines the strategic, cultural, socioeconomic, and geographical differences that influence smart city projects worldwide. It analyzes the geographical influence and internal drivers of smart city projects, focusing on Europe, North America, Latin America, and East and South Asia. This analysis offers insights into diverse approaches to smart city development, encompassing top-down and bottom-up approaches. By examining various perspectives and experiences from smart city initiatives worldwide, this chapter sheds light on the challenges and opportunities associated with implementing smart city strategies in different contexts.

Keywords: Geographic patterns; smart city implementation; organizational drivers; strategic approaches; cultural differences; socioeconomic drivers; top-down approaches; bottom-up approaches

4.1 Setting Smart City Implementation

The implementation of smart city projects is a complex process that involves many strategic and geographic factors. These factors can vary significantly from one country to another, and they can influence the success and trajectories of smart city projects (Christofi, Iaia, Marchesani, & Masciarelli, 2021;

The Global Smart City, 55–74

Published under exclusive licence by Emerald Publishing Limited
doi:10.1108/978-1-83797-575-420231004

Marchesani, Iaia, Masciarelli, & Christofi, 2022; Shelton, Zook, & Wiig, 2015). This chapter provides an overview of the strategic and geographic approaches that influence the implementation of smart city projects globally. Specifically, this chapter will discuss the role of citizen-centric, technology-centric, and decision-making processes in the implementation of smart cities by also examining the strategic, cultural, socioeconomic, and geographical differences that affect the implementation of smart city projects worldwide.

In the context of smart city implementation, the concepts of culture and organizational practices play a critical role. These concepts can significantly influence the success or failure of smart city projects. In this line, Hofstede's cultural dimensions theory could be a framework that provides insights into the cultural and organizational differences that affect the implementation of smart city projects worldwide (Hofstede, 2011).

Hofstede's cultural dimensions theory identifies six dimensions of culture that can influence the behavior and attitudes of individuals and organizations. These dimensions include power distance, individualism–collectivism, masculinity–femininity, uncertainty avoidance, long-term orientation, and indulgence-restraint (Beugelsdijk, Kostova, & Roth, 2017; Hofstede, 2011). The dimensions can help explain the cultural differences that affect the implementation of smart city projects worldwide. For example, in countries with high power distance, there may be a greater emphasis on hierarchical decision-making processes, while in countries with low power distance, there may be a greater emphasis on participatory decision-making processes.

The implementation of smart city projects is influenced by a range of strategic, cultural, socioeconomic, and geographical factors and contributes to the advancement and strategic process into the smart city trajectories (Christofi et al., 2021; Hall & Hubbard, 1998; Marchesani et al., 2022; Salom-Carrasco, Pitarch-Garrido, & Sales-Ten, 2017).

Strategic differences refer to the different approaches that cities take in implementing smart city projects. Specifically, citizen-centric, technology-centric, and decision-making processes are three strategic approaches that can influence the implementation of smart city projects (Marchesani et al., 2022). Citizen-centric smart city projects aim to improve the quality of life of citizens by providing them with better services and amenities. Technology-centric smart city projects aim to improve the efficiency and effectiveness of city services and infrastructure. Effective decision-making processes involve the participation and engagement of all stakeholders, including citizens, city officials, and private sector partners.

In this line, Hofstede's cultural dimensions theory could be a view to get insights into the cultural and organizational differences that affect the implementation of smart city projects. Based on the understanding of the cultural and organizational differences that affect the implementation of smart city projects worldwide, this section briefly discusses the three strategic processes that can influence the implementation of smart city projects. Each of these approaches plays a crucial role in the development and success of smart city projects, varying based on the inherent characteristics and desired goals planned in each city.

4.1.1 Citizen-Centric Process

The citizen-centric process emphasizes the importance of focusing on the needs, aspirations, and well-being of the citizens in the design, planning, and implementation of smart city projects. This approach recognizes that the ultimate goal of smart cities is to improve the quality of life for their inhabitants, and therefore, it is essential to involve the citizens in the decision-making process actively (Jungwoo Lee & Lee, 2014; Nikki Han & Kim, 2021).

In this process, the local government and other stakeholders actively engage with the citizens to understand their needs, preferences, and opinions. This can be achieved through various channels such as public consultations, surveys, workshops, online platforms, and social media (Pereira, Parycek, Falco, & Kleinhans, 2018; Reinartz, Wiegand, & Imschloss, 2019). The insights gained from these interactions are then used to inform the design and implementation of smart city projects, ensuring that they are tailored to the specific needs of the local community (Jungwoo Lee & Lee, 2014).

The citizen-centric approach not only promotes a sense of ownership and inclusiveness among the citizens but also helps to ensure the long-term success and sustainability of smart city projects, as they are more likely to address the actual needs of the community and receive widespread support.

Lessons from the City: Amsterdam, Holland

The city of Amsterdam in the Netherlands has implemented a citizen-centric approach to smart city development through its Amsterdam Smart City initiative. This initiative involves the collaboration of various stakeholders, including citizens, businesses, knowledge institutions, and government agencies, to co-create and implement smart city solutions that address the city's challenges and improve the quality of life of its citizens (Hajer & Bröer, 2020).

One example of a successful citizen-centric smart city project in Amsterdam is the "City-zen" project. This project aims to make Amsterdam's city center energy-neutral by 2040, by involving citizens in the development and implementation of energy-saving solutions. The project includes various elements, such as smart grids, energy storage, and energy-efficient buildings, and it involves the participation of citizens through workshops, co-creation sessions, and pilot projects.

(Dobbelsteen et al., 2020)

Through this citizen-centric approach, Amsterdam has been able to involve citizens in the development and implementation of smart city solutions, resulting in solutions that are better suited to the needs and preferences of the citizens. This approach has also enabled Amsterdam to create a more sustainable and livable city, while fostering a sense of community engagement and ownership among its citizens.

Today, the citizen-centric approach is a crucial element in successful smart city projects (especially in European and American countries), emphasizing the active involvement of citizens in decision-making. By prioritizing the needs, aspirations, and well-being of the local community, this approach ensures that smart city initiatives are tailored to their specific requirements, resulting in enhanced quality of life and long-term sustainability. Through diverse engagement channels like consultations and online platforms, valuable insights are gathered and utilized to design and implement projects, fostering a sense of ownership and inclusiveness among citizens. Overall, the citizen-centric approach signifies a shift in urban governance, emphasizing the importance of cocreating smart city projects. By embracing this approach, cities can harness technology's transformative potential while ensuring equitable distribution of benefits that align with citizens' actual needs and aspirations.

4.1.2 Technology-Centric Process

The technology-centric process focuses on the deployment and integration of advanced technologies to enable the efficient and effective management of urban systems, services, and resources. This approach views technology as a key enabler for achieving the goals of smart cities, such as improving the efficiency of public services, reducing environmental impacts, and enhancing the overall quality of life for citizens (Christofi et al., 2021; Pittaway & Montazemi, 2020; Thatcher, Burns, & Dalton, 2021).

In the technology-centric process, the emphasis is on leveraging cutting-edge technologies, such as the Internet of Things (IoT), big data analytics, artificial intelligence (AI), and blockchain, to support the monitoring, analysis, and optimization of various urban processes (Allam & Dhunny, 2019; Chatterjee, Kar, & Gupta, 2018; Marchesani, 2022). These technologies can be deployed across different domains, such as transportation, energy, water, waste management, and public safety, to create a more efficient, sustainable, and livable urban environment (Linde, Sjödin, Parida, & Wincent, 2021).

While the technology-centric approach can deliver significant benefits in terms of improved efficiency and sustainability, it is important to ensure that the adoption of new technologies is done in a way that is sensitive to the needs and preferences of the citizens, as well as the local context and culture (Hollands, 2015; Kummitha, 2018). This can help to avoid potential issues related to privacy, data security, and social exclusion.

One example of a successful technology-centric smart city project is the City Brain initiative in Hangzhou, China. The initiative aims to use AI and big data analytics to improve the efficiency of city services, such as transportation, public safety, and emergency response. The initiative involves the use of various technologies, such as sensors, cameras, and data analytics, to collect and analyze data on city services and infrastructure. These data are used to develop and implement smart city solutions that improve the efficiency and effectiveness of city services and infrastructure (Liu, Yang, & Nie, 2023; Yun & Lee, 2019).

4.1.3 Decision-Making Process

The decision-making process in smart city implementation refers to the governance structures, policies, and practices that guide the planning, design, and execution of smart city projects (Gil-Garcia, Zhang, & Puron-Cid, 2016). This process plays a crucial role in determining the success and sustainability of smart city initiatives, as it shapes the strategic direction, resource allocation, and stakeholder engagement efforts.

The decision-making process can vary significantly across different cities and regions, depending on factors such as the political system, cultural norms, and administrative structures. Some cities may adopt a more centralized, top-down approach, where decisions are made primarily by the central government or a designated smart city agency. In contrast, others may opt for a more decentralized, bottom-up approach, where decision-making authority is distributed among various local stakeholders, such as municipalities, businesses, and citizens (Bonakdar & Audirac, n.d.; Chan, 2012; Leonelli, Marchesani, & Masciarelli, 2022).

Regardless of the specific approach, effective decision-making in smart city implementation requires a high degree of collaboration, coordination, and transparency among all relevant stakeholders (Abid, Marchesani, Ceci, Masciarelli, & Ahmad, 2022; De Guimarães, Severo, Felix Júnior, Da Costa, & Salmoria, 2020). This can help to ensure that the various smart city projects are aligned with the overarching strategic vision, as well as the specific needs and priorities of the local community.

One example of a successful decision-making process in smart city implementation is the Smart Dubai initiative in the United Arab (Noori, de Jong, Janssen, Schraven, & Hoppe, 2021). The initiative aims to use technology to improve the efficiency and effectiveness of city services, such as transportation, public safety, and health care. The initiative involves a comprehensive decision-making process that involves the participation and engagement of all stakeholders. The process includes the identification of the goals and objectives of smart city projects, the assessment of the risks and benefits of smart city solutions, and the selection of the most appropriate smart city solutions (Khan, Woo, Nam, & Chathoth, 2017).

4.2 Cultural and Socioeconomic Drivers

The development of smart cities is driven by a range of cultural and socioeconomic factors. These factors can influence the design and implementation of smart city projects, and can have a significant impact on the success or failure of these projects (Marchesani et al., 2022; Vinod Kumar, 2020). This section will explore the cultural and socioeconomic drivers of smart city development, and how these drivers influence the trajectory of smart city development.

Cultural drivers refer to the values, beliefs, and attitudes that shape the way people think about and approach smart city development (Hofstede, 2011). These drivers can have a significant impact on the design and implementation of smart city projects.

One cultural driver of smart city development is undoubtedly the concept of trust in government included in the social capital of certain regions and counties (Ooms, Caniëls, Roijakkers, & Cobben, 2020). The level of trust that citizens have in their government can significantly impact the design and implementation of smart city projects. For example, if citizens have high levels of trust in their government's decision-making processes, they may be more willing to accept the collection and use of personal data for smart city projects (Barbi & Mattioli, 2019; Ooms et al., 2020). On the other hand, if citizens have low levels of trust in their government's decision-making processes, they may be more skeptical of smart city projects that involve the collection and use of personal data (Kiss & Muha, 2018).

Another cultural driver is the emphasis on social value and community participation (Laursen, Masciarelli, & Prencipe, 2012; Laursen, Masciarelli, & Reichstein, 2016). Smart city projects that prioritize social value aim to improve the well-being and quality of life for all citizens. This approach recognizes the importance of actively involving the community in the decision-making process, ensuring their needs and aspirations are considered.

Moreover, also the local urbanism is connected to the social, cultural, and economic aspects of urban life. Cities are seen as complex systems that are shaped by a range of factors, including culture, history, and geography (Chang, Jou, & Chung, 2021; Yang, 2020). Urbanism can provide insights into the unique challenges and opportunities that cities face, and can inform the design and implementation of smart city projects.

This cultural driver can have significant implications for the development of smart cities. If citizens do not trust their government to make decisions that are in their best interest, they may be less likely to participate in smart city initiatives, which can limit the effectiveness of these projects. Conversely, if citizens have high levels of trust in their government, they may be more willing to participate in smart city initiatives, which can help to ensure the success of these projects (Borrás & Edler, 2020; Nam & Pardo, 2011a).

However, while cultural drivers are important, it is also crucial to consider the socioeconomic drivers of a nation in the design and implementation of smart city projects.

Socioeconomic drivers refer to the economic and social factors that shape the development of smart cities. These drivers can have a significant impact on the resources that are available for smart city development, as well as the priorities and goals of city governments (Noori, de Jong, & Hoppe, 2020; Sharifi, Khavarian-Garmsir, & Kummitha, 2021).

One socioeconomic driver of smart city development is economic growth. Economic growth can provide cities with the resources that are necessary to invest in smart city projects, such as infrastructure, technology, and human capital (Chatterjee & Kar, 2015; Khatoun & Zeadally, 2016). Smart city projects that are designed to promote economic growth can help to attract new businesses and industries, create new jobs, and increase the overall prosperity of the city.

Another socioeconomic driver of smart city development is social equity. Social equity refers to the principle of fairness and justice in society. Smart city projects that are designed with social equity in mind can help to address issues such as poverty, inequality, and social exclusion (Bonakdar & Audirac, n.d.; Martin, Evans, & Karvonen, 2018). These projects can provide access to essential services and amenities for all members of the community, regardless of their socioeconomic status.

The cultural and socioeconomic drivers of smart city development can have a significant impact on the trajectory of smart city development. These drivers can influence the priorities, goals, and strategies of city governments, as well as the resources that are available for smart city development.

For example, a city that places a high value on sustainability may prioritize smart city projects that are designed to reduce the environmental impact of urbanization. These projects may include the development of renewable energy sources, the promotion of sustainable transportation options, and the implementation of green building practices (Abid et al., 2022).

Similarly, a city that places a high value on social equity may prioritize smart city projects that are designed to address issues such as poverty, inequality, and social exclusion. These projects may include the development of affordable housing, the provision of essential services and amenities, and the promotion of inclusive economic growth (Shelton et al., 2015).

In addition to influencing the priorities and goals of smart city development, cultural and socioeconomic drivers can also influence the trajectory of smart city development. For example, a city that is experiencing rapid economic growth may prioritize smart city projects that are designed to support this growth, such as the development of technology and innovation hubs (Bakici, Almirall, & Wareham, 2013). On the other hand, a city that is experiencing economic decline may prioritize smart city projects that are designed to address this decline, such as the redevelopment of underutilized or abandoned urban areas (Irazábal & Jirón, 2021; Shearmur & Doloreux, 2021).

Thus, it assumes that cultural and socioeconomic drivers of smart city development play a critical role in shaping the design and implementation of smart city projects. These drivers can influence the priorities, goals, and strategies of city governments, as well as the resources that are available for smart city development (Neirotti, De Marco, Cagliano, Mangano, & Scorrano, 2014). By understanding the cultural and socioeconomic drivers of smart city development, city governments can design and implement smart city projects that are better suited to the unique challenges and opportunities of their cities.

4.3 Geographical Influence on Smart City Management

The location of a city in the world can have a significant influence on its approach to smart city management. As anticipated in the previous sections, different countries have unique environmental, social, cultural, and economic characteristics that shape the development objectives and strategies of cities (Christofi et al., 2021; Neirotti et al., 2014). This section explores how geographical location influences smart city management.

The geographical location of a city in the world plays a significant role in shaping its approach to smart city management. Smart cities are designed to leverage technological advancements to improve the quality of life for citizens, enhance sustainability, and foster economic growth (Caragliu & Del Bo, 2021). However, the way smart city projects are managed varies significantly depending on the geographical location of a city.

One factor that can influence smart city management is the political and economic context of a city as anticipated in the previous sections. In centralized political systems, decision-making may be concentrated in the hands of a few individuals, which can limit citizen participation in the planning and implementation of smart city projects. In contrast, more democratic political systems may prioritize citizen engagement and participation in the development of smart city projects (Nesti & Graziano, 2020; Sancino & Hudson, 2020).

The economic context of a city can also shape smart city management. In cities with strong economies, smart city projects may be designed to attract investment and foster innovation, while in cities with weaker economies, smart city projects may be more focused on improving the quality of life for citizens and providing basic services (Abid et al., 2022; Camboim, Zawislak, & Pufal, 2019).

Another factor that can influence smart city management, as anticipated, is the cultural context of a city. Different countries may have different attitudes toward technology and the role it plays in society. Cities may need to consider cultural norms and practices when designing and implementing smart city projects. For example, in cities with a culture of privacy, the collection and use of personal data may be viewed with suspicion, which could impact the implementation of certain smart city technologies (Capdevila & Zarlenga, 2015).

The geographical location of a city can also influence the approach to smart city management in terms of citizen participation (Vidiasova & Cronemberger, 2020). In some regions, there may be a tradition of centralized decision-making, which can lead to a lack of citizen participation in the planning and implementation of smart city projects. In contrast, cities in other regions may prioritize citizen participation and collaboration in the development of smart city projects.

For example, in Asia, there is often a tradition of centralized decision-making (top–down approach as discussed in the next sessions) which can lead to a lack of citizen participation in the planning and implementation of smart city projects. In contrast, cities in Europe and North America may prioritize citizen participation and collaboration in the development of smart city projects.

Thus, it assumes that the geographical location of a city plays a significant role in shaping its approach to smart city management. Smart city projects must be designed with consideration for the unique characteristics of a city, including its political, economic, and cultural context (Boschma, 2005; Christofi et al., 2021; Datta, 2015; Ibrahim, El-Zaart, & Adams, 2018). Understanding these differences is important for policymakers and practitioners who are involved in smart city projects, as it can help them to design and implement projects that are appropriate for the local context and responsive to the needs of citizens. The next section will explore the geographical distribution and internal drivers of smart city projects in more detail.

4.4 Geographical Distribution and Internal Drivers of Smart City Projects

This section explores the geographical distribution of smart city projects and their internal drivers. Building on a systematic review of the literature and the current academic debate about smart city management (Marchesani et al., 2022), this section identifies three different approaches to the smart city strategies: the global-promotion or competitive approach of cities located in Europe and North America, the emerging approach of cities in Latin America, and the top-down approach of cities in East and South Asia. This section aims to highlight how different inputs in the development of cities lead to substantial and well-defined outputs, which turn into different smart city trajectories linked to the geographical reference area.

4.4.1 Europe and North America

Cities located in Europe and North America compete globally and base their strategies on a capitalist perspective to attract innovation, tourism, and human and financial capital (Marchesani, 2022; K. A. Paskaleva, 2009; Taylor Buck & While, 2017). The smart evolution in these areas is driven by the technological and smart transition in services and processes that aim to improve the development of cities by placing the citizen and users at the center of cities' strategies (Christofi et al., 2021). The current need to be sustainable and efficient in medium-long term development leads cities to start and invest in this transformation process. This evolution has impacted both internal and external perspectives and aims to evolve the static concept of cities toward a more comprehensive and global vision (Marchesani, Masciarelli, & Quang, 2022; Snieska & Zykiene, 2015). A global vision in cities development makes the city attractive to citizens and stakeholders and enhances its status worldwide, useful for attracting human capital, talents, and companies to compete locally and internationally.

Lessons from the City: Barcelona, Spain

Barcelona has been recognized as a leader in smart city initiatives, with projects such as the Smart City Expo World Congress and the

Barcelona Urban Lab (Bibri & Krogstie, 2020; Camboim et al., 2019; Global Data Thematic; Grimaldi & Fernandez, 2017; Global Data Thematic Research, 2020). The city has implemented a number of smart technologies, such as smart lighting, smart parking, and a smart grid. These technologies aim to improve the quality of life for citizens and make the city more sustainable and efficient. Barcelona has also invested in innovation, attracting startups and companies to the city and positioning itself as a hub for technology and entrepreneurship (Bakici et al., 2013). Barcelona's leadership in smart city initiatives is a testament to the city's commitment to building a sustainable future for its citizens. By implementing smart technologies and investing in innovation, Barcelona has not only improved the quality of life of its citizens but has also established itself as a global leader in the field (Bakici et al., 2013; Capdevila & Zarlenga, 2015). This, in turn, has attracted businesses, investors, and entrepreneurs to the city, contributing to its economic growth and development. As other cities look to adopt smart city solutions, Barcelona's success provides an excellent example of how technology can be leveraged to create more livable, sustainable, and competitive urban environments.

4.4.2 Latin America

In recent years, several Latin American countries have launched smart city projects aimed at improving their social and urban environments. These initiatives have been developed to tackle internal problems related to urban efficiency and citizen well-being. For example, some cities have implemented smart transportation systems to alleviate traffic congestion, while others have focused on creating more sustainable and energy-efficient environment as well as human-based policies for its users and citizens (Goodspeed, 2015; Irazábal & Jirón, 2021; Jirón, Imilán, Lange, & Mansilla, 2021).

To achieve these goals, many Latin American cities have sought to learn from the experiences of North American and European smart city projects. This has led to the establishment of "twinning relationships" between these cities, whereby they share knowledge and expertise to improve their own urban trajectories. By leveraging these international partnerships, Latin American cities are seeking to enhance their position within the regional ecosystem and promote their status as innovators in the field of urban development (Almirall et al., 2016; Christofi et al., 2021).

At the heart of these smart city initiatives is a desire to put the citizen at the center of the urban renewal project. By focusing on improving the quality of life for their residents, these cities hope to address a range of social and economic challenges, from poverty and inequality to environmental degradation and public health. Through their smart city strategies, Latin American countries are laying the groundwork for a more sustainable and equitable future, both for their own citizens and for the global community as a whole (Jirón et al., 2021).

> **Lessons from the City**: Medellín, Colombia
>
> Medellín has undergone a significant transformation in recent years, with smart city initiatives playing a key role in this process. The city has implemented several projects aimed at improving urban efficiency and citizen well-being. For instance, the Metrocable transportation system connects residents in marginalized areas to the city center, while the City of Life project aims to transform the city's underutilized spaces into vibrant public areas (Bocarejo et al., 2014; Corburn, Asari, Pérez Jamarillo, & Gaviria, 2020). Additionally, the city has incorporated innovative technologies such as smart traffic lights that use sensors to adjust traffic flow in real-time, reducing congestion and emissions. The city has also launched a mobile app that provides citizens with real-time information on public transportation schedules and routes, making it easier for them to navigate the city. Medellín has also focused on promoting its position in the regional ecosystem by hosting events like the Smart City Expo and partnering with other cities in the region to share best practices and knowledge. This emerging approach has encouraged other Latin American cities to pursue similar projects to those in Europe and North America with the aim of solving existing urban problems. The city has collaborated with other European cities, such as Barcelona, to develop joint smart city projects, creating a network of knowledge-sharing and innovation.
>
> (Almirall et al., 2016; Cohen, Almirall, & Chesbrough, 2016)

4.4.3 East and South Asia

Cities in East and South Asia have been at the forefront of embracing smart city technologies to manage their development trajectories. These cities prioritize the use of big data and other control tools to improve their internal management, rather than focusing on international promotion. Their emphasis on socio-technical advancement and user connectivity and monitoring has helped to create more efficient and livable urban spaces (Zhu, Li, & Feng, 2019). For example, smart transportation systems and real-time traffic management tools have helped to reduce congestion and improve commuting times for residents (Yan, Liu, & Tseng, 2020).

However, as with any technological advancement, there are also concerns around privacy and surveillance. Some critics worry that the intense monitoring and data collection associated with smart city technologies could lead to increased government control and potential violations of individual freedoms (Christofi et al., 2021, p. 975). Despite these concerns, the smart cities of East and South Asia continue to push the boundaries of what's possible with technology and urban development. As these cities continue to evolve and innovate, it will be interesting to see how they balance the benefits of smart city technologies with the need to protect individual privacy and freedom (Kiss & Muha, 2018).

Lessons from the City: Seul, Korea

One example of a smart city technology being used in East Asia is the Seoul Smart Mobility System. This system uses big data and real-time traffic information to manage transportation flows in the city, reducing congestion and improving commuting times for residents. The system also includes features like smart parking solutions and an integrated public transportation network, making it easier for residents to get around the city. However, there are also concerns around the potential privacy implications of such a system. Critics worry that the extensive monitoring and data collection required by the system could lead to government surveillance and potential violations of individual privacy. Despite these concerns, the Seoul Smart Mobility System has been successful in making the city more efficient and improving the quality of life for residents (Lee, Yang, Lee, Cho, & Oh, 2016). It serves as an example of how smart city technologies can be used to address urban challenges, while also highlighting the need to consider the potential privacy implications of such technologies.

These different approaches to smart city projects are influenced by geographical context and development objectives. They are also linked to different drivers and outcomes, which can be summarized in Figure 3 (Fig. 3).

Thus, this section highlights the importance of understanding the geographical context and internal drivers of smart city projects. By recognizing the diversity of cities and their trajectories in morphologically, culturally, politically, and economically different countries, we can gain a deeper understanding of the approaches and objectives of smart city projects, and the drivers that shape them.

The last section explores the two different government approaches that emerged in smart city project implementation. Specifically, it will examine the top-down and bottom-up approaches and their respective strengths and weaknesses. By examining these different approaches, we can finally gain a clear overview of the various drivers and outcomes of smart city projects, and the potential trade-offs and challenges associated with each approach.

4.5 Government Approach in Smart City Projects

To conclude this chapter is also important to understand and consider the government's approach to smart city projects as the rise of smart city projects around the world has led to a growing interest in the role of government in shaping the implementation and outcomes of these initiatives. As we have seen in the previous section, the geographical context and internal drivers of smart city projects play a

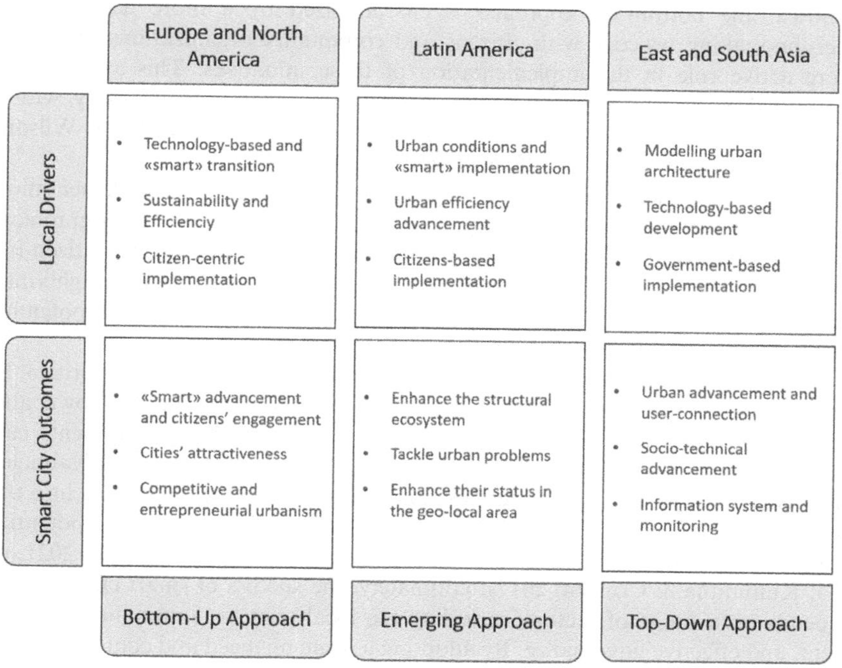

Fig. 3. Drivers and Outcomes of Smart City Projects. *Source*: Created by author.

critical role in shaping their approaches and objectives and become mandatory also examine the role of government in smart city projects.

The role of government in smart city projects is crucial, as it shapes the implementation and outcomes of these initiatives. Government policies, regulations, and funding can significantly impact the success of smart city projects (Pittaway & Montazemi, 2020). Furthermore, the local government is often responsible for coordinating various stakeholders, including private companies, community organizations, and citizens (Kummitha & Crutzen, 2017). Specifically, effective governance is essential for the implementation of smart city projects, as it ensures that resources are allocated efficiently, risks are managed, and benefits are maximized. Government involvement can also help to promote transparency and accountability, which are critical for building trust with citizens and stakeholders.

Two distinct approaches have emerged in the implementation of smart city projects: top-down and bottom-up. The top-down approach is characterized by a centralized decision-making process, with government agencies and private companies leading the implementation of these initiatives. This approach is typically driven by a desire to promote economic growth and attract foreign investment, with a focus on large-scale infrastructure and technology projects. In contrast, the bottom-up approach is characterized by a more decentralized decision-making process, with citizens and community organizations playing a more active role in the implementation of these initiatives. This approach is typically driven by a desire to enhance citizen well-being and sustainability, with a focus on grassroots initiatives and community engagement (Breslin & Wilson, 2015; Capdevila & Zarlenga, 2015; Irazábal & Jirón, 2021).

As previously mentioned, the role of government in shaping the implementation and outcomes of these initiatives cannot be overstated, as well as the geographical location and urban environment that may drive either a governance-led or citizen-led approach (Trencher, 2019). By examining these approaches, we can gain insights into the factors that influence the success of smart city projects, as well as the potential trade-offs and challenges associated with each approach.

Thus, it assumes that the role of government in smart city projects is critical in shaping the approaches and outcomes of these initiatives. The top-down and bottom-up approaches represent two distinct ways in which governments can shape the implementation of smart city projects, each with their own advantages and disadvantages. By examining these approaches, we can gain insights into the factors that influence the success of smart city projects, and the potential trade-offs and challenges associated with each approach (Christofi et al., 2021, p. 975; Kummitha & Crutzen, 2017). Ultimately, the success of smart city projects depends on a range of factors, including the local context, community engagement, and effective governance. By adopting a more nuanced and context-specific approach to smart city projects, we can create more livable, sustainable, and equitable urban environments for users and stakeholders.

4.5.1 Top-Down Approach

The top-down approach is characterized by a centralized decision-making process led by government agencies and private companies. One of the main advantages of the top–down approach is that it allows for the rapid implementation of smart city projects. Decisions are made quickly, and resources are readily available, which enables cities to complete ambitious and complex projects that may have been difficult to achieve through a more decentralized approach (Capdevila & Zarlenga, 2015; Irazábal & Jirón, 2021). Additionally, the top-down approach can be advantageous in cities where there is a lack of community engagement or where there is a need for a centralized decision-making process (Zhu et al., 2019).

However, despite its advantages, the top-down approach also has some disadvantages. One of the main drawbacks is that it may lead to a lack of community engagement and participation. Citizens may feel excluded from the decision-making process, which can lead to a lack of buy-in from the local community and can ultimately undermine the success of the project (Cardullo & Kitchin, 2019). Additionally, the top-down approach may not be suitable for all cities, as it may not take into account the unique cultural, socioeconomic, and political factors that shape the local context.

Cities that have adopted the top-down approach to smart city projects have seen varying degrees of success. For example, in Singapore, the government-led Smart Nation initiative discussed before has resulted in the implementation of several smart city projects, such as the Smart Traffic Management System, which uses data analytics to optimize traffic flow. Similarly, in Shanghai, the city's Smart City Development Plan has focused on improving public services and infrastructure through the use of technology (Odendaal, 2021). However, the success of these projects has not been without criticism, as some have argued that they prioritize economic growth and technological advancement over citizen well-being and sustainability lending to a wide spectrum of critics concerning this approach (especially from the European and North American countries).

One criticism of the top-down approach is that it can be overly focused on technology and infrastructure, without taking into account the needs and perspectives of citizens (Datta, 2015; Marchesani, 2022). This can lead to a lack of community engagement and participation, as well as a potential lack of accountability and transparency in decision-making. Additionally, the top–down approach may not be suitable for all cities, as it may not take into account the unique cultural, socioeconomic, and political factors that shape the local context (Datta & Odendaal, 2019; Hollands, 2015). Critics argue that a more participatory and inclusive approach is needed to ensure the success and sustainability of smart city projects.

Lessons from the City: The Top-Down Approach in China

In China, the government has taken a top-down approach to smart city projects, with a focus on promoting economic growth and technological innovation. Hangzhou, a city in eastern China, has implemented the City Brain project, which uses artificial intelligence and data analytics to manage traffic flow and improve public safety. The project collects data from various sources, such as traffic cameras and GPS sensors, and uses this information to optimize traffic flow in real-time. By analyzing traffic patterns and predicting congestion, the system can adjust traffic signals and reroute drivers to reduce congestion and improve safety.

(Caprotti & Liu, 2022)

Similarly, in the city of Shenzhen, the government has launched the Smart City Blueprint, which aims to create a more efficient and sustainable city through the use of technology. The initiative includes several projects, such as the Smart Transportation Project, which uses data analytics to optimize public transportation routes and reduce congestion. The city has also implemented a smart waste management system, which uses sensors and data analytics to optimize waste collection and disposal.

(Große-Bley & Kostka, 2021)

> However, despite the potential results and the effective benefits in terms of city management and efficiency, there are still concerns about privacy and surveillance, as the City Brain project and other smart city projects in China rely on the collection and analysis of vast amounts of data. Critics argue that more transparency and accountability are needed to ensure that citizens' rights and privacy are protected.
>
> (Datta, 2015; Kiss & Muha, 2018)

It assumes that the top-down approach to smart city projects can be beneficial in certain circumstances, particularly in cities where there is a need for rapid implementation or centralized decision-making. However, it is important for governments to carefully consider their approach and ensure that they engage with citizens and stakeholders throughout the implementation process to ensure the success and sustainability of the project.

4.5.2 Bottom-Up Approach

The second type of approach to implementing smart city projects is the bottom-up approach, which is characterized by a more decentralized decision-making process with citizens and community organizations playing a more active role in the implementation of these initiatives.

One of the main advantages of the bottom-up approach is that it promotes citizen engagement and participation. Citizens and community organizations are actively involved in the decision-making process, which can lead to a greater sense of ownership and investment in the project (Cardullo & Kitchin, 2019; Jungwoo Lee & Lee, 2014). Additionally, the bottom-up approach can be advantageous in cities where there is a need for a more tailored and context-specific approach, as it takes into account the unique cultural, socioeconomic, and political factors that shape the local context (Kummitha & Crutzen, 2017).

Despite its advantages, the bottom-up approach also has some disadvantages. One of the main drawbacks is that it can be slower and more difficult to implement than the top-down approach. Additionally, the bottom-up approach may not be suitable for all cities, particularly those with limited resources or where there is a lack of community engagement (Nesti & Graziano, 2020).

In this approach, citizens and community organizations play an active role in the decision-making process, providing input and feedback on the initiatives that will directly affect their lives. Social media has become an important tool for engaging citizens and promoting community involvement, allowing for real-time feedback and communication (Pereira et al., 2018; Pittaway & Montazemi, 2020). For example, in Chapter 2, it was highlighted that social media has the potential to increase transparency, accountability, and trust between citizens and government, which can ultimately lead to more effective and sustainable smart city projects.

Over the years, cities that have adopted the bottom–up approach to smart city projects have seen varying degrees of success. For example, in Copenhagen, the city's Climate Plan was developed through a bottom-up approach, as well as in Milan, the city's Sharing Cities program focuses on promoting sustainable and equitable urban development through the use of technology and community engagement (as we can see in the examples below).

However, even regarding the bottom-up approach, there are criticisms that (as with the top-down approach), contribute to the current challenge and the need to find a common path in the implementation of smart city projects. For example, in cities where there is a lack of community engagement, the bottom-up approach may not be effective because there may not be enough interest or participation to drive the project forward (Gagliardi et al., 2017). Additionally, in cities with limited resources, it may be challenging to implement a more decentralized approach that relies heavily on community involvement, as it may require more time, effort, and resources to engage citizens and community organizations. In these cases, a more top-down approach may be necessary to ensure that the project is completed effectively and efficiently.

However, despite the natural criticisms and ongoing debate surrounding smart city projects, there are many interesting examples such as Copenhagen and Milan that demonstrate the actual benefits of this approach, as previously mentioned.

Lessons from the City: The Bottom-Up Approach in Copenhagen

The Green Roofs program in Copenhagen is a key initiative included in the city's Climate Plan, which was developed through a bottom-up approach with active participation from citizens and stakeholders (Baykal, 2012; Versini, Gires, Tchiguirinskaia, & Schertzer, 2020). The program aims to promote the use of rooftop gardens as a means of improving air quality and reducing the urban heat island effect, which is a phenomenon where urban areas experience higher temperatures than surrounding rural areas due to the absorption and retention of heat by buildings and pavement. The program has been successful in engaging citizens and promoting sustainable urban development, as it provides a practical and tangible solution to reducing the city's carbon footprint (Versini et al., 2020). The use of green roofs helps to absorb pollutants and reduce greenhouse gas emissions, as well as providing a natural cooling effect that can reduce energy consumption in buildings. Additionally, the gardens can help to reduce stormwater runoff, which can help to mitigate the risk of flooding in the city. The Green Roofs program has also been successful in promoting community engagement and participation. Citizens and community organizations have been involved in the design and implementation of the program, which has helped to build a sense of ownership and investment in the initiative.

(Baykal, 2012)

Overall, the Green Roofs program in Copenhagen serves as an excellent example of how a bottom-up approach to smart city projects can yield positive results for both citizens and the environment. By engaging citizens and stakeholders in the decision-making process and promoting sustainable urban development, the city has been able to create a more livable, resilient, and environmentally friendly urban environment.

Lessons from the City: The Bottom-Up Approach in Milan

In Milan, the city's Sharing Cities program focuses on promoting sustainable and equitable urban development through the use of technology and community engagement (Bernardi & Diamantini, 2018; Salvia & Morello, 2020). Specifically, the program includes a range of initiatives, such as the Sharing Mobility platform, which allows citizens to share electric vehicles and bicycles. By promoting the sharing economy, the program helps to reduce carbon emissions, increase mobility options for citizens, and promote more sustainable and equitable urban development.

The Sharing Mobility platform has been successful in engaging citizens and promoting sustainable urban development, as it provides a practical and effective solution to reducing the city's carbon footprint. The platform allows citizens to easily access and share electric vehicles and bicycles, which can help to reduce the number of cars on the road and promote more sustainable transportation options (Salvia & Morello, 2020). Additionally, the program has been successful in promoting community engagement and participation, as citizens and community organizations have been involved in the design and implementation of the initiative (Bernardi & Diamantini, 2018). By promoting sustainable urban development through the use of technology and community engagement, the program has been able to create a more livable, resilient, and environmentally friendly urban environment in a vibrant and overpopulated city like Milan.

It is assumed that the bottom-up approach to smart city projects can be beneficial in certain circumstances, especially in countries and regions where government and citizens have a close relationship, and both entities influence each other without imposing their views but rather through interconnected positions (Christofi et al., 2021; Marchesani et al., 2022). Therefore, it is essential for cities to carefully consider their approach and engage with citizens and stakeholders throughout the implementation process to ensure project success and sustainability. The continuous connection between citizens and governments in this case not only affects the livability of the city but also its competitiveness and attractiveness to external users who find an urban environment that puts them at the center of its dynamics and development (Abid et al., 2022; Vanolo, 2014). This not only guarantees short-term efficiency but also long-term prosperity for the cities.

4.6 Implications, Perspectives, and Urban Trajectories

The implementation of smart cities has far-reaching implications for various aspects of urban life. This chapter has provided a forward-looking perspective on the development of smart cities within the urban context, considering different types of implementation and approaches. By examining the geographic patterns and drivers discussed in previous sections, we can gain valuable insights into the complex dynamics and challenges associated with smart city development.

Firstly, citizen engagement and inclusivity emerge as key factors in successful smart city projects. Actively involving citizens in decision-making processes and incorporating their feedback ensures that smart initiatives are aligned with the needs and aspirations of the community (Jungwoo Lee & Lee, 2014; Lnenicka et al., 2022; Trencher, 2019). By fostering a sense of ownership and participation, cities can create a shared vision for their smart city journey. Citizen-centric approaches empower individuals and communities to contribute to the design, implementation, and monitoring of smart solutions. This assumption has also emerged within the consideration of case studies, which have provided practical examples and best practices to highlight successful instances of citizen engagement and inclusive smart city projects (Capdevila & Zarlenga, 2015; Christofi et al., 2021; Leitheiser & Follmann, 2020).

Secondly, technological foundations and digital infrastructure play a crucial role in the realization of smart cities. Robust connectivity, efficient data management systems, and interoperability between different urban systems are recognized as essential for the seamless integration of smart solutions. Cities must invest in reliable and secure technological frameworks to support the efficient implementation of smart initiatives (Caprotti & Liu, 2022; Carter et al., 2020; Harrison et al., 2010; Massana, Pous, Burgas, Melendez, & Colomer, 2017). By establishing open data platforms, cities can enable innovation and collaboration, creating an ecosystem of smart applications and services. This chapter has discussed the investments required to build the necessary digital infrastructure and ensure its scalability and adaptability and the different geographical approach to achieve the internal objectives.

Furthermore, data-driven decision-making is a fundamental aspect of smart city governance. The availability of vast amounts of real-time data allows policymakers and urban planners to make informed decisions and implement evidence-based policies (Berrone, Ricart, & Carrasco, 2016; Caragliu & Del Bo, 2019; De Guimarães et al., 2020; Setijadi et al., 2019). The utilization of data analytics and AI enables cities to optimize resource allocation, enhance urban management, and improve the overall quality of life for residents. However, data privacy and security remain critical considerations, requiring robust governance frameworks and ethical guidelines to protect individual rights and ensure responsible data usage. Addressing social disparities and inequalities is another critical implication of smart city development (Giarda & Moroni, 2018; Leonelli et al., 2022; Reddick, Enriquez, Harris, & Sharma, 2020).

While technological advancements can bring numerous benefits, there is a risk of exacerbating existing inequalities if access to smart technologies and services is unevenly distributed. Ensuring equitable access and bridging the digital divide are imperative to create inclusive and socially sustainable smart cities (Albino, Berardi, & Dangelico, 2015; Bibri & Krogstie, 2020; Chatterjee & Kar, 2015; Vanolo, 2014).

In examining these implications, perspectives, and urban trajectories, this chapter provides valuable insights into the multifaceted nature of smart city development. It offers practical guidance for users, citizens, public managers, and policymakers involved in the planning and implementation of smart city initiatives. The journey toward smart cities requires a holistic and context-specific approach that considers the unique characteristics, challenges, and strengths of each city. Tailoring smart solutions to specific needs and contexts ensures their relevance and effectiveness. In doing so, balancing top-down guidance with bottom-up participation is vital in smart city projects. While top-down approaches provide an overarching vision and strategic direction, grassroots initiatives and local participation empower communities and foster innovation. A collaborative governance model that encourages partnerships between the government, private sector, academia, and citizens can lead to more inclusive and responsive smart city development (Capdevila & Zarlenga, 2015; Irazábal & Jirón, 2021; Leitheiser & Follmann, 2020; Mariani, Di Felice, & Mura, 2016; Tang, Jayakar, Feng, Zhang, & Peng, 2019). This chapter has highlighted successful examples of collaborative governance and explores the benefits of multi-stakeholder engagement.

Therefore, it is assumed that the implications, perspectives, and urban trajectories associated with smart city development are vast and multifaceted, and closely tied to the approach, geographical distribution, and socioeconomic context in which the city is located. As we have seen throughout the earlier parts of this book, smart cities have the potential to revolutionize urban living by providing innovative solutions to complex challenges. By placing citizen engagement at the forefront, harnessing the power of technology, promoting inclusivity, and fostering collaboration, cities can unlock the full potential of smart solutions and create sustainable, resilient, and livable urban environments. The journey toward smart cities necessitates a holistic and context-specific approach that strikes a balance between top-down guidance and bottom-up participation (Marchesani et al., 2022). As cities navigate the intricacies and seize the opportunities presented by the smart city era, the pursuit of sustainable urban development remains of utmost importance. By drawing lessons from past experiences, leveraging emerging technologies, and embracing inclusive governance models, cities can shape their future and enhance the overall quality of life for all residents.

Part Two
Urban, Economic, and Competitive Outcomes of the Smart City Projects

Chapter 5

Urban Environment in the Smart City

Abstract

This chapter aims to provide a comprehensive understanding of the urban outcomes of smart city projects, focusing on their primary objectives. The first objective is to facilitate the management and flow of information, data, and resources to enhance resource efficiency, sustainability, and the quality of life for citizens and stakeholders. This chapter offers insights into the urban objectives of smart city projects within the local ecosystem, with a specific emphasis on digital and key urban outcomes. It provides an overview of the digital outcomes, including the advancement of digital systems for safety and urban monitoring, the provision of customized digital services, and the promotion of citizen engagement through digital platforms. This chapter also evaluates the environmental outcomes of smart city projects, such as improved quality of life, increased urban efficiency, and contributions to a sustainable environment. To provide a well-rounded understanding, interviews with policymakers and city managers, as well as case studies from cities like London, Medellin, Helsinki, Singapore, Girona, and San Diego, are incorporated. Furthermore, this chapter incorporates data and findings from top-tier international journals to provide a clear understanding of the impact of smart cities on the local ecosystem.

Keywords: Urban environment; smart city; efficiency; sustainability; quality of life; digital outcomes; environmental outcomes; citizen engagement; local ecosystem

5.1 Smart City Objectives and Expected Outcomes

As expounded upon in the first part of the book, the internal factors comprising management and practices constitute significant drivers in shaping the objectives of smart cities. These objectives, in turn, hold paramount importance in realizing the wider aims established by the European Union and other global entities concerning urban development, sustainability, and the welfare of citizens. Nevertheless, it remains imperative to acknowledge the results and the obstacles

The Global Smart City, 77–98

Published under exclusive licence by Emerald Publishing Limited
doi:10.1108/978-1-83797-575-420231005

associated with accomplishing these objectives and appraising their advantages. This section aims to concentrate on the tangible benefits of smart cities and provide an overview of the themes to be expounded upon in the following sections.

The primary objectives of smart cities are multifaceted and encompass various aspects of urban life. These objectives include improving the quality of life for citizens, enhancing urban efficiency, promoting environmental sustainability, and fostering economic growth (Albino, Berardi, & Dangelico, 2015; Vanolo, 2014). To achieve these objectives, smart cities leverage advanced digital systems, innovative technologies, and data-driven solutions that can transform urban environments into more resource-efficient and sustainable ecosystems (Caragliu & Del Bo, 2019; Kummitha & Crutzen, 2017; Linde, Sjödin, Parida, & Wincent, 2021).

The European Union has recognized the importance of smart cities and their potential in addressing pressing urban challenges, such as climate change, resource depletion, and population growth. As a result, the European Union has set forth several policy initiatives and funding programs aimed at promoting smart city development across its member states.[1] Some of these initiatives include the European Innovation Partnership on Smart Cities and Communities (EIP-SCC) and the Horizon 2020 program, which emphasize the need for innovative solutions to urban challenges and the importance of collaboration between public and private sectors, academia, and civil society in achieving these objectives.

Despite the increasing emphasis on smart city development and the potential benefits they offer, there are several challenges in realizing these objectives. One of the main hurdles is the difficulty in evaluating the benefits of smart city initiatives. Due to the complex nature of urban environments and the interdependencies between various systems and stakeholders, assessing the impact of smart city projects is not a straightforward task (Nikki Han & Kim, 2021). Furthermore, the lack of standardized metrics and methodologies for evaluating smart city outcomes makes it challenging to compare and benchmark the performance of different cities (Wiig, 2015).

Another challenge is the need for significant investments in infrastructure, technology, and human resources to support the development and implementation of smart city solutions. This can be particularly difficult for cities with limited financial resources or those facing other pressing issues, such as aging infrastructure or social inequalities (Chatterjee & Kar, 2015; Zhao, Fashola, Olarewaju, & Onwumere, 2021). Additionally, the rapid pace of technological advancements and the need for continuous adaptation to new innovations can create further challenges in implementing and maintaining smart city solutions.

Despite these challenges, smart cities offer tangible benefits that can significantly improve the urban environment and citizens' lives. Some of these benefits include:

---

[1]EPRS: European Parliamentary Research Service. Belgium. Retrieved from https://policycommons.net/artifacts/1339578/mapping-smart-cities-in-the-eu/1949353/

(1) Enhanced public safety and security through advanced monitoring and surveillance systems, which can help detect and prevent crime, traffic accidents, and other incidents (Goodspeed, 2015).
(2) Improved traffic management and transportation efficiency through intelligent transport systems, which can optimize traffic flow, reduce congestion, and minimize emissions (Benevolo, Dameri, & D'Auria, 2016).
(3) Increased energy efficiency and reduced carbon footprint through smart grid technologies, which can optimize energy consumption and promote the use of renewable energy sources (Angelidou, Politis, Panori, Bakratsas, & Fellnhofer, 2022).
(4) Enhanced waste management and resource efficiency through smart waste collection and recycling systems, which can optimize waste collection routes and promote recycling initiatives (Chauhan, Jakhar, & Chauhan, 2021).
(5) Improved access to public services and information through digital platforms and mobile applications, which can facilitate communication between citizens and government agencies and promote civic engagement (Yun & Lee, 2019).

These tangible benefits not only contribute to a more efficient and sustainable urban environment but also have a direct impact on citizens' well-being and quality of life. For instance, improved air quality resulting from reduced emissions can lead to better public health and reduced health-care costs (Gagliardi et al., 2017; Linde et al., 2021). Enhanced public safety can foster a sense of security and trust within communities, while increased access to public services can promote greater social inclusion and equity.

The upcoming sections of this chapter aim to provide a deeper exploration of the (i) digital and (ii) environmental outcomes associated with smart cities, with the intent of offering an all-encompassing comprehension of the initial objectives and final results of smart city initiatives. More specifically, Section 5.2 will delve into various case studies and leverage insights from policymakers, city managers, and academic research to present a comprehensive outlook on the influence of smart cities on the local ecosystem.

5.2 Digital Outcomes in the Urban Environment

The integration of technology into urban environments has brought about a significant transformation in the way cities function. The advent of smart city initiatives has led to a host of benefits that have impacted not only the cities but also their residents. This discussion aims to delve into the impact of digital outcomes associated with smart cities on various areas such as public safety, accessibility to public services, urban mobility, citizen participation, innovation, and urban and economic development. Specifically, the role of technology in transforming urban environments has been crucial in achieving a wide range of benefits for both cities and their residents (Barham & Daim, 2020; Christofi, Iaia, Marchesani, & Masciarelli, 2021).

Technology has been instrumental in making cities more efficient, safe, and accessible. Through the use of sensors, data analytics, and renewable energy sources, smart cities have reduced their carbon footprint while ensuring a consistent supply of energy (Cheng, Awan, Ahmad, & Tan, 2021). In addition to energy efficiency, technology has also improved public safety and transportation. The integration of CCTV cameras, sensors, and data analytics has enabled smart cities to monitor public spaces, identify potential safety threats, and respond quickly to emergencies (B. N. Silva, Khan, & Han, 2018). The implementation of intelligent transportation systems has made traffic management more efficient, resulting in reduced congestion and improved traffic flow. Moreover, technology has played a significant role in making public services more accessible to citizens. Through digital platforms and mobile apps, citizens can now access real-time information on various public services such as health care, transportation, and waste management. These platforms have made it easier for citizens to access public services efficiently, thereby increasing their overall satisfaction with their city's administration (S. Kumar, Mookerjee, & Shubham, 2018; Simonofski, Hertoghe, Steegmans, Snoeck, & Wautelet, 2021).

Another benefit of technology in smart cities is the improvement of urban mobility and reduced traffic congestion. Smart cities have implemented various measures such as bike-sharing systems, pedestrian zones, and electric vehicle charging stations, promoting sustainable transportation. These initiatives have significantly reduced traffic congestion, making it easier for citizens to navigate through the city while reducing their carbon footprint (Benevolo, Dameri, & D'Auria, 2016).

Finally, technology has enabled cities to become more transparent and participatory. Through online platforms and social media, citizens can now voice their opinions and concerns about their city's administration, policies, and services (Nam & Pardo, 2014). This has resulted in increased citizen engagement, leading to better-informed policy decisions and a stronger sense of community.

Although technology has brought significant benefits to smart cities, it has also raised concerns about privacy, security, and the potential for technological malfunctions. As cities become more digitally integrated, the risk of cyber-attacks and data breaches becomes more significant (Kiss & Muha, 2018; Z. Li & Liao, 2018). Furthermore, technological malfunctions could result in system failures that may have severe consequences for citizens' safety and well-being.

However, the benefits of technological and digital implementation are clear and effective, and the role of technology in smart cities cannot be overstated as digital implementation in the urban environments has led to a wide range of benefits. The objective of this section is to explain the actual outcomes of the digitalization process between cities and citizens, with a focus on safety and monitoring, customized and tailored services, and citizens' engagement. These outcomes contribute to shaping the relationship between users and the city, helping to achieve the objectives of efficiency, quality of life, and sustainability that will be highlighted in Section 5.2.1.

5.2.1 Safety and Monitoring of the Urban Environment

The safety and monitoring of the urban environment is a critical aspect of smart city initiatives, and technology has played a significant role in achieving this goal. With the help of various digital tools and technologies, smart cities have been able to monitor and respond to potential safety threats in real-time.

One of the most common tools used for safety and monitoring in smart cities is the CCTV camera (Hashem et al., 2016; Rahman et al., 2019). These cameras are strategically placed in public spaces and monitored by trained personnel to detect and respond to potential safety threats quickly. For instance, London has implemented a complex and advanced system to monitor passenger safety in the cities' transports and prevent crime.

> **Lessons from the City**: London, United Kingdom
>
> Transport for London (TfL) has implemented several measures to improve public safety and prevent crime in the city's public transport system. One such measure is the installation of over 12,000 CCTV cameras across the public transport network, including the London Underground, bus network, and Docklands Light Railway (Song, Cao, Han, & Hickman, 2019). These cameras are strategically placed at stations, platforms, and on-board vehicles to monitor passenger behavior and detect potential safety threats. The footage captured by these cameras is monitored in real-time by TfL's security team, who are able to identify and respond to incidents quickly. In addition, the cameras act as a deterrent to potential offenders, reducing the likelihood of crime occurring in the first place.
>
> TfL has also implemented other digital tools to improve passenger safety, such as the Safer Travel App. The app allows passengers to report incidents of crime or anti-social behavior directly to TfL's security team. Passengers can also use the app to receive real-time alerts about disruptions to the transport network and to plan their journeys accordingly (https://www.london.gov.uk).
>
> Overall, TfL's use of digital tools and technologies has significantly improved the safety and security of the city's public transport system. The integration of CCTV cameras and other monitoring tools has enabled TfL to detect and respond to safety threats quickly and effectively, while also deterring potential offenders. These measures have contributed to improving the safety and reliability of London's public transport system.

Apart from CCTV cameras, sensors and data analytics are also used for safety and monitoring in smart cities. For example, sensors can detect hazardous conditions such as gas leaks or fires, and alert the authorities immediately. Data analytics can also be used to identify patterns and trends that may indicate potential safety threats, enabling the authorities to respond quickly (Allam & Jones, 2020).

Another significant aspect of safety and monitoring in smart cities is emergency response systems. In case of an emergency, citizens can use mobile apps or digital platforms to contact emergency services quickly. In this sense, in Singapore, the government has launched a mobile app called SGSecure that enables citizens to report potential security threats and receive real-time updates during emergencies.

> **Lessons from the City**: Singapore
>
> In Singapore, the government has developed a mobile app called SGSecure, which serves as an all-in-one platform to provide citizens with real-time updates during emergencies and to report potential security threats. This app enables citizens to receive alerts and notifications from relevant authorities in the event of an emergency, such as terrorist attacks or natural disasters. Additionally, SGSecure allows citizens to report suspicious activities or persons by submitting photos or videos, which can then be used by the authorities to investigate and prevent security threats.
>
> The app also provides guidance on emergency response procedures and offers training modules to prepare citizens for various emergency situations. Furthermore, SGSecure serves as a communication tool for residents to connect with their local community groups and stay informed about community safety initiatives.
>
> (Chia, 2017)
>
> Overall, the SGSecure app is an effective example of how digital tools can be used to improve public safety and security in smart cities. By leveraging technology to facilitate timely and accurate information sharing and citizen engagement, cities like Singapore can enhance their emergency response capabilities and ensure the safety and well-being of their citizens.

Smart cities have also implemented digital tools to improve the safety of pedestrians and cyclists contributing to the advance of the security and quality of life in the city. For example, smart pedestrian crossings use sensors to detect pedestrians and cyclists and adjust traffic lights to provide them with safe crossing times. These crossings use various sensors, such as infrared sensors or cameras, to detect the presence of pedestrians and cyclists waiting to cross the street. Once the sensors detect the presence of a pedestrian or cyclist, they send a signal to the traffic lights at the intersection. The traffic lights then adjust their timing to provide a safe crossing time for the pedestrian or cyclist. Smart pedestrian crossings are designed to reduce the risk of accidents involving pedestrians and cyclists, who are often more vulnerable than motorists (Carter et al., 2020). By providing a safe crossing time and minimizing the need for pedestrians and cyclists to rush across the street, these crossings can help to prevent accidents and improve the overall safety of the urban environment.

In conclusion, it is assumed that the safety and monitoring of the urban environment is one of the most critical aspects of smart city initiatives. With the help of various digital tools and technologies, smart cities have been able to monitor and respond to potential safety threats in real-time. CCTV cameras, sensors, data analytics, emergency response systems, and intelligent transportation systems are just some of the digital tools used for safety and monitoring in smart cities. By ensuring the safety of their citizens, smart cities are not only improving their quality of life but also attracting businesses and investments, making them more competitive in the global market.

5.2.2 Customized and Tailored Services

In the digitalization process and technology-centric urban implementation, customized and tailored services have become increasingly important in the context of smart cities. These services can range from transportation and health care to waste management and public safety. By leveraging technology, cities can gather data on individual citizens and use it to personalize services, thereby increasing efficiency and improving citizen satisfaction.

The use of digital technologies in these practices offers several benefits to both citizens and cities. Firstly, they improve citizen satisfaction and overall quality of life by providing services that are tailored to individual needs (Trencher, 2019). This leads to greater citizen engagement and participation in the city's activities. Secondly, personalized services can help cities become more efficient in their delivery of services, which can result in cost savings and better resource allocation (Khatoun & Zeadally, 2016) as discussed in the subsequent section related to the urban outcomes.

To address these two points, some of the key technological solutions include the use of Internet of Things (IoT) devices and sensors for data collection. These devices can collect data on citizen behavior and preferences, which can be used to personalize services such as traffic management and waste management (Rahman et al., 2019; Ullah, Qayyum, Thaheem, Al-Turjman, & Sepasgozar, 2021). By analyzing these data, cities can gain insights into the needs and preferences of their citizens. Another solution is the use of big data analytics for processing and analyzing large amounts of data. This allows cities to identify patterns and trends in citizen behavior, which can be used to personalize services. Artificial intelligence (AI) and machine learning (ML) are also being used to personalize services based on individual needs (S. Kumar et al., 2018). AI algorithms can be used to predict traffic congestion and suggest alternative routes to individual drivers. Additionally, ML algorithms can be used to analyze data on citizen behavior and preferences, and make recommendations for personalized services based on these data. Finally, cloud computing is being used to store and share data across multiple devices and platforms, enabling cities to access and analyze data in real-time (Hämäläinen, 2020). This can improve the delivery of personalized services, as cities can quickly respond to changing citizen needs and preferences.

> **Lessons from the City**: Helsinki, Finland
>
> Helsinki has implemented a digital platform called 'MyHelsinki' that offers personalized recommendations to both visitors and residents on events, restaurants, and other activities based on their preferences and behavior (Hämäläinen, 2020). The platform uses data collected from various sources, including social media and user feedback, to provide personalized recommendations to users. For example, if a user has shown an interest in music events, the platform will suggest concerts and other music-related activities.
>
> In addition to providing personalized recommendations, the platform also encourages citizen engagement. Users can provide feedback and suggestions on the recommendations they receive, which helps to improve the quality and relevance of the recommendations. This two-way communication also fosters a sense of ownership and responsibility among citizens, as they feel they are actively contributing to the development of the city's services.
>
> The 'MyHelsinki' platform has been successful in improving citizen engagement and satisfaction in the city. It has received positive feedback from users and has helped to promote Helsinki as a tourist destination by providing personalized and relevant recommendations to visitors. The platform's success has also inspired other cities to develop similar platforms and services to improve citizen engagement and satisfaction (https://www.myhelsinki.fi).

Moreover, the outcomes of utilizing these technological solutions for personalized services in smart cities are numerous. By tailoring services to individual needs, cities can improve citizen satisfaction and increase efficiency. For example, personalized traffic management can reduce congestion and travel time, while personalized waste management can reduce environmental impact and increase recycling rates (Marchesani, 2022; Spicer, Goodman, & Olmstead, 2021).

However, there are also challenges to be addressed, such as privacy concerns and the need for equitable access to technology. Their implementation can be challenging due to privacy concerns, the digital divide, and the need for fair and unbiased implementation by local governments. It is crucial to address these challenges to ensure the successful implementation of personalized services in smart cities and provide equitable access to all citizens, leading to more efficient and effective services (Ellison, Bannister, Lee, & Haleem, 2021). Thus, the engagement between citizens and the city is also crucial in ensuring the successful implementation of personalized services, as it fosters a sense of ownership and responsibility among citizens and municipalities, leading to more effective and sustainable services.

5.2.3 Citizens' Engagement

Citizen engagement in cities refers to the involvement of citizens in decision-making processes that affect their lives and their environment. In a smart city, the use of technology can facilitate this engagement, enabling citizens to participate more effectively in the design and management of their urban environment (Mossberger, Wu, & Crawford, 2013; Nam & Pardo, 2011a). In the smart city context, this engagement could be seen as form of democratic participation that allows citizens to express their opinions and concerns about the issues that affect them, and it can take many forms, including participation in public meetings, online forums, surveys, and social media platforms.

Citizen engagement is crucial for smart cities, as it allows for the cocreation of urban environments (Ooms, Caniëls, Roijakkers, & Cobben, 2020). By involving citizens in the design and management of the city, smart cities can ensure that urban policies and services are tailored to the needs and preferences of their citizens. Citizen engagement can also promote transparency and accountability in the decision-making process, enhancing trust between citizens and the city. In this process, digital implementation has assumed a central role. For example, data analysis techniques can be used to analyze the data collected from citizen data collection technologies. This analysis can provide insights into the needs and preferences of citizens, enabling smart cities to design and deliver services that are tailored to their citizen (Tura & Ojanen, 2022). At the same time, communication and information technologies, such as social media platforms and mobile apps, can facilitate communication between citizens and the city (Gagliardi et al., 2017). These technologies can be used to inform citizens about urban policies and services, as well as to gather feedback and opinions from citizens.

The use of engagement technologies can have a significant impact on the efficiency and effectiveness of public policies. By involving citizens in the design and management of urban policies and services, smart cities can ensure that they are tailored to the needs and preferences of their citizens (Bernardi & Diamantini, 2018).

Engagement technologies can also have a positive impact on citizen participation in urban activities. By providing citizens with the tools to participate in decision-making processes and cocreate their urban environment, smart cities can promote a sense of ownership and belonging among their citizens (Global Data Thematic Research, 2020; Ooms et al., 2020). In this line, cocreation represents an interesting outcome of the citizen's engagement as it refers to a collaborative process in which citizens and other stakeholders work together to create or improve a product, service, or environment. In the context of smart cities, cocreation involves citizens and local governments collaborating to design and implement solutions that address urban challenges and improve the quality of life for residents. Thus, engagement technologies can facilitate cocreation by providing citizens with easy access to information and tools that enable them to participate in decision-making processes and contribute to the development of their city (Pereira, Parycek, Falco, & Kleinhans, 2018). For example, a city may use an online platform to gather feedback and suggestions from citizens on urban planning initiatives or to crowdsource ideas for new services or amenities. By providing citizens with a direct channel to participate in decision-making processes and cocreate their urban environment, smart cities can foster a sense of ownership and belonging among their citizens (Löfgren & Webster, 2020).

Lessons from the City: Medellin, Colombia

In Medellin, Colombia, the city government has developed a program called "Participation and Digital Citizenship" to engage citizens in the decision-making process (Echeverry Tamayo, 2019; Irazábal & Jirón, 2021). The program is facilitated by a digital platform where citizens can submit their proposals, ideas and projects related to different aspects of urban development, such as public space design, mobility, and social inclusion. The platform also allows citizens to vote on proposals submitted by other citizens.

This approach has helped to improve citizen participation and involvement in the city's decision-making processes. By leveraging technology to connect citizens with the city government, Medellin has been able to tap into the collective knowledge and creativity of its citizens, leading to innovative and effective solutions to urban challenges.

One notable project that emerged from this program is the "Cinturon Verde" (Green Belt) initiative, which involves converting former garbage dumps into parks and green spaces (Echeverry Tamayo, 2019). This project was proposed by a citizen and won the most votes from other citizens on the digital platform. The city government recognized the proposal and allocated funds to make it a reality, resulting in a significant improvement to the quality of life for local residents.

However, despite these positive and effective advantages, the use of citizen data collection technologies raises concerns about privacy and security. Smart cities must ensure that citizens' data are protected and that they are informed about how their data are being used. The collection of personal data can be perceived as intrusive and may raise concerns about surveillance (Löfgren & Webster, 2020). Therefore, it is important to establish clear guidelines on how data is collected, stored, and used. Additionally, measures such as data encryption, secure servers, and access controls can help ensure the security of citizen data.

Another challenge of citizen engagement technologies is the issue of digital exclusion. Not all citizens have access to the necessary technologies or skills to participate in digital engagement platforms. This can lead to a digital divide, where certain groups are left out of the engagement process. It is important to ensure that citizen engagement technologies are accessible to all citizens, regardless of their socioeconomic status or technological skills (Datta, 2015).

Finally, the use of citizen data in engagement technologies can raise ethical and legal issues. For example, there may be concerns about the use of personal data for commercial purposes or the use of algorithms that may reinforce bias or discrimination. It is important to ensure that engagement technologies are developed and used in an ethical and transparent manner (Vith, Oberg, Höllerer, & Meyer, 2019). Additionally, legal frameworks should be established to regulate the use of citizen data in engagement technologies.

Thus, it is assumed that citizen engagement is a crucial aspect of smart cities, and technology plays an important role in facilitating citizen engagement. The use of technologies such as data collection, data analysis, communication, and participation platforms can help cities better understand the needs and preferences of citizens, and create more effective and efficient policies and services. The ultimate goal of smart city initiatives is to improve environmental outcomes, with a particular focus on enhancing life quality and equality, promoting urban efficiency, and creating a sustainable urban environment. Citizen engagement technologies play a critical role in achieving these goals by empowering citizens to participate in decision-making processes, cocreate their urban environment, and provide valuable insights that can inform the development of more effective policies and services. The impact of citizen engagement on these environmental outcomes will be further discussed in Section 5.3.

5.3 Environmental Outcomes of Smart City

In the preceding sections of this book, a comprehensive analysis was conducted to explore the motivations and strategies that underlie the environmental outcomes of smart cities. Building upon this foundational understanding and the digital outcomes previously considered, this section will delve into a more detailed evaluation of the real-world impacts that smart city projects have had on the urban environment.

This section discusses the specific, tangible outcomes of smart city initiatives and how they are shaping the development trajectories of cities. The development of smart cities is driven by a range of factors, including technological advancements, changes in management practices, and various drivers such as population growth, urbanization, and environmental challenges. These factors have led to the development of a range of characteristics and dimensions that contribute to shaping smart city projects (Christofi et al., 2021, p. 972). This construct was discussed in earlier sections of the book, where the motivations and strategies behind the environmental outcomes of smart cities were analyzed. By orchestrating the six dimensions of smart cities with digital technology implementation and public management, cities can build a more sustainable and resilient future for their residents (Albino et al., 2015; Linde et al., 2021). Each dimension has a specific and tangible impact on the urban environment, and addressing them with smart city initiatives can lead to improved efficiency, sustainability, and well-being for urban residents. This will ultimately contribute to the creation of more livable and desirable cities that are better equipped to address the challenges of the future.

Specifically, the mobility dimension has enabled efficient movement of people and goods, reducing traffic congestion and emissions in urban areas. This has resulted in improved air quality and reduced carbon footprint, contributing to a more sustainable and livable urban environment (Benevolo, Dameri, & D'Auria, 2016). In the environment dimension, the use of renewable energy sources and the reduction of waste have led to a significant decrease in greenhouse gas emissions and a more sustainable use of resources. Improved air and water quality, as well as the creation of green spaces and natural habitats, have also contributed to enhancing the quality of life for urban residents (Dobbelsteen et al., 2020). The people dimension of smart cities has resulted in healthier lifestyles, increased social inclusion, and a stronger sense of community engagement. This has been achieved through the promotion of physical activity, access to health-care services, and the creation of public spaces for social interaction, resulting in improved well-being for urban residents (Marchesani, 2022). In the living dimension, sustainable urban design and the use of green infrastructure have led to more resilient built environments and reduced environmental impact of buildings. This has resulted in the creation of more livable and sustainable urban spaces, with increased access to green areas and reduced energy consumption (Abid, Marchesani, Ceci, Masciarelli, & Ahmad, 2022).

Finally, the governance dimension has enabled effective and equitable implementation of smart city initiatives, ensuring that environmental goals are prioritized in urban development, and that the benefits of smart city initiatives are shared equitably among urban residents. This has been achieved through the promotion of transparency, citizen engagement, and collaboration between different stakeholders, resulting in a more democratic and participatory approach to urban governance (Gagliardi et al., 2017).

It is assumed that the successful implementation of smart city initiatives requires the effective development and orchestration of all dimensions. To guarantee the smart cities' successful implementation and promotion, institutions such as the European Union and the United Nations have played and still play a critical role in providing the necessary support to ensure that sustainability and social inclusion objectives are at the forefront of the smart city agenda.[2] The digital and technology aspects are important, but they must be integrated with other dimensions such as mobility, environment, people, and living, to create a holistic and sustainable

---

[2]EPRS: European Parliamentary Research Service. Belgium. Retrieved from https://policycommons.net/artifacts/1339578/mapping-smart-cities-in-the-eu/1949353/

approach to smart city development. Specifically, the European Union, the United Nations, and their respective initiatives, such as the 17 Sustainable Development Goals (SDGs) and Agenda 2030, have played a critical role in promoting the environmental outcomes of smart cities (Vanolo, 2014).

In this line, concrete actions have been taken, and significant results have been achieved in many cities around the world. For example, the European Union has supported numerous smart city projects through funding programs and knowledge-sharing platforms, contributing to the successful implementation of sustainable technologies and practices in urban environments (K. A. Paskaleva, 2011). Cities like Amsterdam, Barcelona, and Copenhagen (as anticipated in the previous examples) have become global leaders in smart city development, showcasing innovative solutions for improving urban sustainability with the final aims to make cities inclusive, safe, resilient, and sustainable, which is directly aligned with the environmental outcomes of smart city strategies.

This section provides a detailed evaluation of the specific and tangible outcomes of smart city initiatives and how they are shaping the development trajectories of cities. The analysis focusing on the three main environmental outcomes inherent in smart cities: improving life quality and equality, enhancing urban efficiency, and promoting a sustainable urban environment.

5.3.1 Life Quality and Equality

Life quality and equality are essential outcomes of smart city projects. These aspects significantly impact the well-being of city dwellers and contribute to the overall success of urban areas. Specifically, several smart city solutions have been implemented to improve the quality of life and promote equality in urban areas. In this sense, smart city implementation and digital technologies in cities have the potential to significantly enhance the accessibility and efficiency of essential services in urban areas, including health care, education, and public transportation (Camboim, Zawislak, & Pufal, 2019; Chauhan et al., 2021). By leveraging data analytics, AI, and the IoT, these technologies can streamline service delivery and improve overall efficiency. For example, by integrating real-time traffic data and intelligent transport systems, smart city technologies can optimize public transportation routes, reducing traffic congestion and improving travel times. This can lead to increased access to employment and education opportunities for residents, particularly those living in marginalized communities (Chatterjee & Kar, 2015; Kar, Ilavarasan, Gupta, Janssen, & Kothari, 2019). In addition, smart city technologies can improve the delivery of health-care services by enabling remote patient monitoring, telemedicine, and predictive analytics. These solutions can help to reduce healthcare costs and improve patient outcomes, particularly for individuals living in underserved or remote areas (Gaur, Scotney, Parr, & McClean, 2015). Furthermore, smart city technologies can enhance the accessibility and quality of education through online learning platforms, digital libraries, and collaborative learning tools (Anttila & Jussila, 2018). These solutions can help to bridge the digital divide and provide equal opportunities for education to all residents, regardless of their socioeconomic background.

Another important benefit of smart city solutions is the reduction of pollution and noise levels. By optimizing public transportation systems, promoting eco-friendly practices, and reducing traffic congestion, smart city technologies can help to reduce air and noise pollution, improving the overall well-being of citizens (Chu, Cheng, & Yu, 2021). This is particularly important in densely populated urban areas where pollution and noise levels can have a significant impact on residents' health and quality of life. Furthermore, smart city solutions can engage citizens in environmental sustainability efforts through education and awareness campaigns, and by providing opportunities for public participation in decision-making processes as seen before. By fostering a sense of community ownership and responsibility, these solutions can promote sustainable behaviors and reduce pollution and advance the sustainable environment in urban areas (Khatoun & Zeadally, 2016).

Finally, smart city projects leverage technology and policy to bring citizens closer to the city, orchestrating various dimensions to offer everything citizens need (Linde et al., 2021). As discussed in Chapter 3, this offering ranges from mobility and health care to a thriving economy, high security and integration, and even catering to people's passions by promoting events and offering a 360-degree service to citizens. This comprehensive approach attracts citizens to the city, making it the perfect container for their lives, and by leveraging technology, data, and policy, smart cities can create a sustainable, livable, and prosperous urban environment that drives economic growth and enhances the well-being of citizens.

While smart cities have the potential to provide numerous benefits for urban residents, it is important to consider potential negative consequences that may arise from their implementation. These consequences include increased living costs, privacy invasion and security threats, and the creation of digital inequalities (Marchesani, Iaia, Masciarelli, & Christofi, 2022; Rahman et al., 2019; S. Y. Tan & Taeihagh, 2020). Specifically, one potential negative consequence of smart city solutions is the increase in living costs. The implementation of these solutions may require significant investments in infrastructure and technology, which may result in higher costs for residents. This could lead to exclusion of low-income individuals from the benefits provided by smart cities, exacerbating existing social and economic inequalities.

Another potential concern is the invasion of privacy and security threats. The use of advanced technologies such as video surveillance and data analysis can raise concerns about the collection and use of personal information. It is important for smart city solutions to implement appropriate privacy and security measures to protect citizens' rights and prevent misuse of information.

Furthermore, smart city solutions may unintentionally create digital divides, resulting in unequal access to technology and services. While smart city technologies can enhance accessibility and efficiency, they may also create inequalities between those who have access to technology and those who do not. This could result in the exclusion of certain groups from the benefits provided by smart cities, further exacerbating existing social and economic inequalities.

To ensure the potential benefits of smart city solutions are fully realized, it is crucial to implement these initiatives with a focus on equity and inclusion. This includes measures to ensure that the benefits of smart cities are accessible to all residents, regardless of their socioeconomic status or access to technology (Batty et al., 2012; Datta & Odendaal, 2019; Vanolo, 2014). Additionally, implementing appropriate privacy and security measures is important to protect citizens' rights and prevent misuse of personal information.

Smart city initiatives have already led to tangible improvements in life quality by offering a more efficient and inclusive urban environment. By leveraging technology and data, these initiatives create a more responsive and adaptive urban environment that meets the needs of all citizens. As citizens experience better living conditions in smart cities, they are more likely to move to and thrive in these urban areas. This leads to increased economic growth, job creation, and investment, creating a virtuous cycle of development and prosperity (Abid et al., 2022; Marchesani et al., 2022; K. A. Paskaleva, 2009; M. Porter, 1998).

Overall, smart city solutions offer numerous benefits to urban areas, and it is crucial to recognize and fully embrace these benefits. By implementing smart city solutions with a focus on equity and inclusion, we can create a more sustainable and equitable urban environment that benefits all residents, and improves the overall well-being of our communities. Below is a practical example of how the smart city project in San Diego has effectively improved the livability and quality of life of its citizens.

Lessons from the City: San Diego, California

One practical example of how a smart city project has improved the quality of life for its citizens is San Diego, California. The city has implemented several smart city solutions, such as LED public lighting and intelligent traffic management, to improve the sustainability and safety of the urban environment (Hollands, 2008; Perry, 2018). San Diego has also implemented several digital solutions to improve the efficiency of public services and communication with citizens. For instance, the city has developed an app that enables citizens to report issues like potholes or broken streetlights directly to the relevant authorities, improving the speed and effectiveness of infrastructure maintenance. The implementation of LED public lighting has resulted in significant energy savings for the city, reducing costs and the environmental impact of lighting. The intelligent traffic management system has also improved traffic flow, reducing congestion and emissions, leading to a more sustainable environment. Additionally, the app developed by San Diego has improved citizen engagement and communication with local authorities, enabling citizens to report issues and receive updates in real-time, leading to more efficient service delivery and a better quality of life for residents.

In summary, the implementation of smart city solutions in San Diego has resulted in significant benefits for its citizens, including improved sustainability, safety, and efficiency of the urban environment, as well as improved citizen engagement and communication with local authorities.

Thus, it is assumed that smart city solutions have the potential to significantly improve life quality and equality in urban areas. It is crucial to continue developing and implementing such solutions while minimizing any potential side effects. Collaboration among governments, nongovernmental organizations, businesses, and citizens is essential in designing and implementing effective smart city solutions that enhance life quality and equality for all residents.

5.3.2 Urban Efficiency

The second urban outcome in smart city implementation is to achieve urban efficiency. This refers to the ability of cities to optimize their resources and services in a way that maximizes their effectiveness while minimizing waste. Specifically, urban efficiency can be defined as the ability of a city to use its resources and services in the most effective way possible. As previously discussed, efficient infrastructures in smart cities extend beyond transportation and energy systems, including public infrastructure such as schools and community centers (Gaur et al., 2015; Linde et al., 2021; Maria Luisa Marsal-Llacuna & Segal, 2017). These infrastructures play a crucial role in the quality of life of citizens, and optimizing their efficiency has led to significant benefits in terms of sustainability, cost savings, and improved services. These benefits have been observed by city management and contribute to shaping and driving smart city initiatives (Marchesani et al., 2022). For example, one of the interviews conducted revealed that:

> We are witnessing remarkable benefits in terms of energy efficiency, especially in public buildings and construction projects approved during the latest National Recovery and Resilience Plan (PNRR). This is because policies not only encourage energy efficiency but also aim to gain energy to be used within the city. Of course, these practices are costly, and without the support of EU policies, they would not have been possible, but the benefits are remarkable. We have managed to significantly lower the city's energy consumption by working on lighting, mobility, services, and infrastructure. Currently, we are among the most virtuous projects in terms of energy efficiency. This inspires us to improve further and continue along this path, which is yielding excellent results.

These benefits are supported by the use of advanced technologies such as the IoT, and data analytics has enabled cities to collect real-time data on energy usage, occupancy, and other factors that can affect the efficiency of public buildings, and this large amount of data has been analyzed to identify areas where improvements can be made, such as adjusting heating and cooling systems or optimizing lighting to reduce energy consumption.

In terms of energy efficiency, the integration of renewable energy sources such as solar panels and wind turbines has also been implemented in public buildings, resulting in reduced reliance on nonrenewable energy sources and lower carbon footprint. In addition, the use of energy storage systems such as batteries has ensured a consistent supply of energy even during periods of low renewable energy production (Angelidou et al., 2022; Neirotti, De Marco, Cagliano, Mangano, & Scorrano, 2014). Data analytics has also been used to optimize the use of public infrastructure in smart cities. For example, data on occupancy levels in community centers and schools have been used to adjust schedules and staffing levels to ensure that facilities are used efficiently. Data on traffic patterns around public buildings have also been used to optimize transportation routes and reduce congestion (Staletić, Labus, Bogdanović, Despotović-Zrakić, & Radenković, 2020).

Lessons from the City: Girona, Catalunya

In the city of Girona, smart city technologies have been implemented to improve energy efficiency and optimize the use of public infrastructure (Maria Lluïsa Marsal-Llacuna, Colomer-Llinàs, Meléndez-Frigola, 2015; Massana, Pous, Burgas, Melendez, & Colomer, 2017). For example, the city has installed smart streetlights that use LED bulbs and are equipped with motion sensors, allowing them to automatically adjust their brightness according to the level of pedestrian and vehicular traffic. This has resulted in significant energy savings for the city.

In addition, the city has implemented a smart irrigation system that uses data from weather sensors and soil moisture sensors to optimize water usage in parks and other public spaces. This has led to a reduction in water consumption and has helped the city to conserve its water resources. Moreover, the city has also implemented a smart parking system that allows drivers to easily find available parking spaces, reducing traffic congestion and carbon emissions. The system uses sensors to detect available parking spots and displays the information on a mobile app, making it easier for drivers to locate a parking spot.

Overall, these innovative solutions in Girona have contributed to the city's goal of becoming more sustainable and efficient, while also improving the quality of life for its residents (https://www.girona.cat/catedrasmartcity/eng/projectes.php).

Efficiency in cities extends beyond just buildings and infrastructure, but also impacts the daily lives of citizens and urban safety. Smart cities recognize this and implement intelligent services designed to improve citizen experience, such as smart parking, intelligent lighting, and waste management. These services provide greater convenience, time savings, and increased safety for citizens (Battarra, Gargiulo, Tremiterra, & Zucaro, 2018; Carter et al., 2020). Moreover, safety is a critical component of urban efficiency, and smart cities utilize advanced technologies such as video surveillance, sensors, and drones to ensure citizen safety. By deploying these technologies, cities can detect and prevent crime in real-time and create a safer environment for their citizens contributing to the life quality previously discussed.

Thus, it is assumed that efficient urbanization is a fundamental goal of smart cities. The results achieved by these cities in terms of urban efficiency have been impressive, with significant benefits for citizens. However, there is still much work to be done to ensure that all cities can achieve urban efficiency. As cities continue to grow and evolve, it is essential that they prioritize urban efficiency as a key component of their development strategies. The benefits of urban efficiency are tangible and significant, and cities that prioritize this goal will undoubtedly reap the rewards.

5.3.3 Sustainable Urban Environment

Finally, sustainable urban development is a fundamental aspect of smart cities, which aims to promote a better quality of life for all citizens. Sustainable urban development encompasses various aspects such as urban planning, energy efficiency, waste management, water resource management, and pollution reduction. These aspects are crucial for achieving a sustainable urban environment and promoting the well-being of citizens. Specifically, the sustainable outcomes have a significant impact on the development of cities, both in terms of urban architecture and sustainable outcomes (Abid et al., 2022; Maria Lluïsa Marsal-Llacuna et al., 2015).

In terms of urban architecture, urban land-use planning involves the allocation and management of land for different purposes, such as residential, commercial, and industrial uses. Smart cities promote sustainable urban land-use planning by encouraging mixed-use development that allows for a diverse range of activities in an urban environment. This can reduce the need for long-distance travel, thereby reducing carbon emissions and congestion. In this urban architecture, transportation planning plays a critical aspect. Smart cities aim to promote sustainable transportation options such as public transit, cycling, and walking, which can reduce the number of cars on the road and thereby reduce air pollution and congestion. In addition, smart cities are implementing innovative transportation solutions such as electric and autonomous vehicles, which can further reduce the carbon footprint of transportation. This urban architecture is optimized to meet current and future climate needs, while also promoting green spaces and paying particular attention to the environment (Contreras & Platania, 2019; Kar et al., 2019; Tiwana, 2014).

In this line, the promotion of green spaces is another important aspect of sustainable urban planning in smart cities. Green spaces such as parks and gardens can provide numerous benefits such as enhanced biodiversity, improved air quality, and reduced urban heat island effects (Carter et al., 2020; Zygiaris, 2013). For example, in Singapore, the city has implemented a green corridor project that connects various green spaces and provides citizens with access to nature. This project not only enhances the quality of life of citizens but also promotes biodiversity and ecosystem services. Overall, sustainable urban planning is a critical aspect of smart cities' efforts to create sustainable urban environments. By promoting mixed-use development, sustainable transportation options, and the promotion of green spaces, smart cities can enhance the quality of life of citizens while minimizing negative impacts on the environment (Maria Lluïsa Marsal-Llacuna et al., 2015).

As previously discussed, in this urban architecture, the environmental outcomes are integrated to contribute to the improvement and sustainability of cities, such as energy, waste management, water management, and pollution reduction. These are timely issues that cities are currently facing and will continue to address in the years to come.

Specifically, in terms of sustainable energy, smart cities promote the use of sustainable energy sources and energy-efficient technologies to reduce the carbon footprint of cities. This includes the use of renewable energy sources such as solar and wind power, energy-efficient buildings, and smart grids (Benevolo, Dameri, & Auria, 2016; Jindal, Kumar, & Singh, 2020). For instance, in Amsterdam, the city has implemented a smart grid system that optimizes the use of renewable energy sources and reduces energy waste. Smart cities also promote sustainable waste management practices such as recycling, waste reduction, and the use of innovative technologies to reduce waste (Dobbelsteen et al., 2020).

Another effect of smart city projects is the efficient use of water resources through the use of innovative technologies and water management practices. This includes the use of water-efficient technologies, rainwater harvesting, and the efficient management of wastewater (Clift et al., 2017; Morgan & Webb, 2020). For example, in Barcelona, the city has implemented a smart water management system that monitors water usage and identifies areas where water can be saved. Last but not least, smart cities promote pollution reduction through the use of innovative technologies and policies (Chauhan et al., 2021). This includes the reduction of greenhouse gas emissions, the monitoring of air and water quality, and the promotion of sustainable transportation. For instance, in Paris, the city has implemented a low-emission zone that restricts the use of high-emission vehicles in certain areas, reducing the level of air pollution:

Lessons from the City: Paris, France

In Paris, the city has implemented a low-emission zone that restricts the access of high-emission vehicles in certain areas through the use of automatic number plate recognition systems. The low-emission zone covers an area of approximately 105 square kilometers and affects around 140,000 vehicles per day. Only vehicles that meet specific emissions standards are allowed to

> enter the designated zone, which is located within the city's urban core. The goal of this initiative is to improve air quality by reducing the level of air pollution caused by vehicular traffic, which is a significant contributor to urban air pollution. This initiative is part of Paris' broader effort to promote sustainable transportation options and reduce the city's carbon footprint, and it has been successful in reducing air pollution levels and improving the health and well-being of citizens living in the affected areas.
>
> (Poulhès & Proulhac, 2021)

Thus, we clearly understand that smart cities are transforming the way cities operate, placing sustainability at the forefront of urban development. As some of the most significant contributors to pollution, cities are taking action to address the pressing issues of climate change and environmental degradation. Through sustainable urban planning, energy efficiency, waste management, water resource management, and pollution reduction, smart cities are creating a better quality of life for citizens while minimizing negative impacts on the environment. This demonstrates a growing recognition of the importance of sustainability in urban development, and the role that smart cities can play in promoting sustainable outcomes.

5.4 Assessing and Evaluating Urban Outcomes

Finally, following the implementation phase and the actual technological and environmental impact, it is imperative to assess these projects and produce a report outlining the benefits they have contributed to the urban environment. Such an evaluation serves not only to account for investments made in smart city projects but also to comprehend the practical utility and cost-benefit analysis derived from these implementations, while simultaneously enabling the creation of useful best practices for the development of analogous projects on a global scale (Batty et al., 2012; Camboim et al., 2019; Cardullo & Kitchin, 2019). Specifically, the evaluation of urban outcomes is a critical aspect of smart city development that involves various stakeholders, including cities, the European Union, and external policies (K. A. Paskaleva, 2011; B. Yuan & Li, 2019).

Of course, as a primary actor, cities play a crucial role in evaluating urban outcomes. Cities can evaluate the outcomes by setting specific parameters and targets for these projects and tracking their progress over time. Additionally, cities can use citizen feedback to evaluate the effectiveness of smart city projects. Citizen feedback can provide valuable insights into the impact of these projects on urban life and help identify areas for improvement (Gagliardi et al., 2017).

This evaluation requires considering the important role of global policies in the development of these projects. For example, the European Union provides funding and support for smart city projects across Europe. In exchange for this support, the European Union requires cities to provide detailed reports on the outcomes of these projects. The European Union uses these reports to evaluate the effectiveness of smart city projects and to identify areas for improvement. Additionally, the European Union provides guidance and best practices for evaluating urban outcomes, helping cities to develop effective evaluation frameworks (Archibugi, Filippetti, & Frenz, 2020).

In this context, the reporting and evaluation of smart city projects is entrusted to the "National and International Project Support Service," which collaborates with political decision-makers and governments. The service is accountable for ensuring that the evaluation process is continuous and demonstrable. For instance, an interesting internal perspective emerged during the interviews regarding this matter:

> Our office not only participates in tenders related to smart city projects, but also follows the calls proposed and submits requests to implement these projects in the city. We also take care of reporting and monitoring the implementation process, which is carried out in accordance with European Union standards (as seen earlier, electronic reporting and sustainability standards for the city and companies involved). Specifically, as a city, we receive part of the funding in the project proposal, which is used for feasibility studies, and then additional funds to apply and develop the project. Over the next 4 years, we must continuously report on the expenses and benefits and ensure that this service (or infrastructure) is active and operational for at least 4 years after the funding. This part is very important because it allows governments to assess the actual need and development of these projects.

In this evaluation, an important role is played by global policies on the development of such projects. The National and International Project Support Service is responsible for evaluating the projects and reporting on their benefits to the urban environment. The service works in collaboration with political decision-makers and governments to ensure that the evaluation process is continuous and demonstrable.

As a consequence, reporting helps to ensure accountability and transparency in smart city projects. By reporting on the outcomes of these projects, cities can demonstrate the effectiveness of their investments and identify areas for improvement. Reporting also provides valuable feedback to citizens and other stakeholders, allowing them to evaluate the impact of smart city projects (Abid et al., 2022; Masik, Sagan, & Scott, 2021). This relationship with users and citizens has become even more interesting due to the recent policies on smart cities in Europe,[3] where it is clearly stated that cities must report to citizens and users.

> Currently, with the recent circular from the European Union on funds related to smart city projects (April, 2023), it has become

---

[3]https://commission.europa.eu/funding-tenders/find-funding/find-calls-funding-topic/culture-and-media_en

> mandatory to report not only to the European Union but also to citizens about the development of these projects. For instance, we must include the European Union logo in these projects, advertise the development and implementation of these projects through cities' social media channels, and be specific and clear in terms of usage and engagement regarding online projects. This means not only clearly communicating access and usage on the portal through open data but also highlighting in a clear manner (which is a new requirement) the use of these funds and the development of smart projects in the city in collaboration with governments and the European Union.

In conclusion, the assessment and evaluation of smart city projects rely heavily on reports on urban outcomes, which play a fundamental role in shaping future investments in smart city projects. These reports offer valuable insights into the effectiveness of specific projects and identify best practices for future implementations. It is essential to prioritize continuous and demonstrable evaluations and reporting of the benefits brought about by these projects, with collaboration from national and international project support services, policymakers, and governments, to ensure transparency, accountability, and the identification of the best practices for smart city development.

However, to truly benefit citizens and address their needs, it is imperative to prioritize citizen engagement and urban needs over mere technological implementation. Evaluating urban outcomes is not just a matter of accountability; it is a crucial step toward creating a smarter, more sustainable, and livable city. By prioritizing citizen engagement and urban needs, smart city projects can make a significant difference in people's lives and pave the way for a sustainable urban future.

Chapter 6

Economic and Business Environment in the Smart City

Abstract

This chapter examines the impact of smart city projects on the economic and business environment, with a specific focus on their potential to foster innovation and entrepreneurship. It highlights how smart cities create an enabling environment for new businesses, start-ups, and promotes gender and green entrepreneurship, while enhancing urban properties and contributing to the overall entrepreneurial ecosystem. The exponential growth of smart city projects worldwide has underscored the critical relationship between smart cities and the entrepreneurial ecosystem for city prosperity and competitiveness. This chapter explores the interplay between smart cities and the economic environment, emphasizing their role as hubs for innovation and business creation. It analyzes the influence of smart city ecosystems and economic factors on local growth and economic outcomes, drawing on international research and factual data. This chapter evaluates the economic outcomes of smart cities, including their impact on the economic environment, business creation, start-ups, innovative and green companies, as well as gender entrepreneurship. This chapter emphasizes the significance of smart city implementation in driving economic advancements within both the urban and economic landscapes. Finally, this chapter concludes by presenting a conceptual framework that synthesizes the insights gained, providing a comprehensive understanding of the economic impact of smart cities and offering a roadmap for further exploration of this topic.

Keywords: Smart city; economic impact; business environment; innovation; entrepreneurship; female entrepreneurship; green economy; new businesses; economic ecosystem

The Global Smart City, 99–115

Published under exclusive licence by Emerald Publishing Limited
doi:10.1108/978-1-83797-575-420231006

6.1 Introduction: The Interaction Between Urban and Business Environment

This chapter aims to contribute to the practical understanding of the relationship between smart cities and the economic environment, and their effects on businesses and entrepreneurship. Within the context of economic geography and smart city literature, scholars suggest that the economic environment in cities, regions, and countries is partly determined by the urban development of the geographical area in which businesses are located. Therefore, smart cities represent an important hub and a prominent context for business purposes. The importance of smart cities in contributing to the economic environment is supported by extensive theoretical discussion, but empirical evidence is still needed due to the relative novelty of the phenomenon (Batabyal & Nijkamp, 2019; Betz, Partridge, & Fallah, 2016; Kummitha, 2018; Muñoz & Cohen, 2016; Vanolo, 2014). However, if the effect on the economic environment is merely a result of smart city implementation, each city should have an equal probability of becoming a prominent economic hub. This probability should be a consequence of smart city transformation and contribute to a well-developed economic environment.

Nevertheless, building on the empirical evidence presented in this chapter, it highlights that the effect could be divergent, as different internal policies consequently impact various economic areas and lead to diverse economic outcomes. For example, over the years, we have seen numerous examples of urban agglomeration and local initiatives in creating entrepreneurial ecosystems based on technology and innovation. Silicon Valley in the United States is a tangible example of such a concentration (Chandler & Saxenian, 1995; English-Lueck & Saveri, 2001). An advanced ecosystem, combined with a well-developed, less-developed, or transitioning economic area, has long been considered a factor that may explain the differences in the economic environment among urban areas and cities (Castelnovo, Morretta, & Vecchi, 2020; Leonelli, Marchesani, & Masciarelli, 2022). However, in considering the smart city context, the urban ecosystem directly influences the entire economic and business environment, affecting its trajectories and guiding its development in a vision that aligns with the objectives and characteristics of smart cities.

This chapter explores the role of smart cities in promoting an advanced and competitive business environment by enhancing the benefits resulting from an equal, efficient, inclusive, digital, and sustainable city (Albino, Berardi, & Dangelico, 2015; Vanolo, 2014). While the research debate on this topic focuses on the role of geographical agglomeration (Black & Henderson, 1999; Glaeser & Gottlieb, 2009; Holmes, Fujita, Krugman, & Venables, 2000; Knudsen, Florida, Gates, & Stolarick, 2007), social capital (Laursen et al., 2012, 2016; Nakano & Washizu, 2021) and firm-level organizational (Bacq & Janssen, 2011; Bikfalvi, Jäger, & Lay, 2014) drivers, this chapter focuses on the smart city environment, suggesting that an advanced urban environment in the geographical area may play a role in determining entrepreneurial opportunities and business creation.

6.2 Smart City Ecosystem and Economic Environment

The number of smart city projects continues to grow exponentially around the world and impact many aspects of our lives, including the economic environment. As a result of this global growth, the relationship between smart cities and the entrepreneurial ecosystem has become fundamental for the prosperity and competitiveness of the city, as the economic environment influences both business opportunities and personal financial prosperity (Adler, Florida, King, & Mellander, 2019; Hollands, 2020; Kummitha, 2019).

In discussing the economic outcomes of smart cities, this chapter focuses on the effect of smart city projects on the economic and business environment, highlighting the interaction between the city and entrepreneurial ecosystems. Today, smart cities offer an advanced environment for businesses, entrepreneurs, and talent to thrive, thus contributing to the economic growth and prosperity of the city (Kummitha & Crutzen, 2019; Marchesani, Masciarelli, & Quang, 2022; Neumann, Matt, Hitz-Gamper, Schmidthuber, & Stürmer, 2019). As such, smart cities have become important hubs for innovation and business creation, attracting a diverse range of companies and individuals from all over the world.

Throughout history, cities have always played a crucial role in the economy. As centers of population and activity, cities provide fertile ground for innovation, creativity, and entrepreneurship (Batabyal & Nijkamp, 2019; Richard Florida, 2002b; Kummitha, 2019; S. Y. Lee, Florida, & Acs, 2004; Marchesani & Masciarelli, 2021). Today, more than half of the world's population lives in cities, and this trend is expected to continue due to disparities between cities and urban areas in services, opportunities, amenities, and job opportunities (Chatterjee & Kar, 2015; Winters, 2011). As a result, cities have become a fundamental crossroads for the future economic landscape, with many actively promoting businesses, attracting headquarters, and supporting start-ups and entrepreneurs.

In particular, smart cities have emerged as hotspots for innovation and entrepreneurship, with advanced infrastructure, technology, and data-driven decision-making (Breslow, 2021; Gagliardi et al., 2017; Linde, Sjödin, Parida, & Wincent, 2021). Furthermore, by offering a conducive environment for businesses to operate and grow, cities have the potential to drive economic growth and prosperity not just for themselves but for the entire region and country (Berrone, Ricart, & Carrasco, 2016). This development is translated into the creation of an urban ecosystem where businesses, universities, policymakers, and entrepreneurs contribute to the development of the city and impact the creation of a prosperous and advanced economic environment.

The urban ecosystem of smart cities refers to a complex system of interconnected physical and digital infrastructures, institutions, and actors collaborating to create an intelligent and sustainable urban environment (Camboim, Zawislak, & Pufal, 2019). This ecosystem encompasses various components, including smart transportation systems, energy-efficient buildings, and digital communication networks, and involves multiple stakeholders such as citizens, businesses, governments, and research institutions (Vanolo, 2014). These actors work together to design and implement innovative solutions to urban challenges, with data and information playing a crucial role in decision-making and resource allocation. The smart city ecosystem is dynamic and constantly evolving, driven by technological advances, changing societal needs, and new opportunities for collaboration and innovation. Its ultimate goal is to create a more livable, sustainable, and resilient urban environment that benefits all stakeholders (Christofi, Iaia, Marchesani, & Masciarelli, 2021; Ooms, Caniëls, Roijakkers, & Cobben, 2020). The success of this ecosystem depends on the ability of its participants to work together in a coordinated and effective manner, creating a virtuous cycle of innovation and prosperity for the city and its residents.

6.3 City as a Hub for Innovation and Business Creation

Cities have nowadays assumed a central role in social and economic geography as a vanguard in urban innovation, catalyzing humans, knowledge, and companies. In the digital era, cities are changing their trajectories following the "smart" revolution, which aims to manage, offer, support, and guarantee proper planning for cities and stakeholders (Camboim et al., 2019). In the transition from cities to smart cities, technology has taken on a pivotal role as a driver and instrument in the current cities' trajectories, orchestrating a pool of actors, needs, structures, policies, and strategies contributing to generating and advancing local ecosystem (Kummitha & Crutzen, 2017; Kummitha, 2019; Mouton & Burns, 2021; Pittaway & Montazemi, 2020; Sharifi, Khavarian-Garmsir, & Kummitha, 2021). Moving toward a more comprehensive concept of a smart city interrelated with innovation, knowledge, entrepreneurship, and stakeholders (Abid, Marchesani, Ceci, Masciarelli, & Ahmad, 2022; Albino et al., 2015; Camboim et al., 2019; Kummitha & Crutzen, 2017; Vanolo, 2014), researchers and academics have long debated about the influence that cities' digital and technological advancement has historically had on attracting and harnessing talent and creating a proper environment suitable for innovative companies (Richard Florida, Adler, & Mellander, 2017; Kummitha & Crutzen, 2019, p. 51; Nieto & Quevedo, 2005). More specifically, digital advancement in the city promotes an innovative environment for companies, which can build on an innovation-oriented ecosystem, and stimulates the creation of new and innovative companies (Adler & Florida, 2021; ****Marchesani, Masciarelli, & Quang, 2022).

A smart city aims to increase its competitiveness through innovation while improving the quality of life of its citizens and offering a technologically advanced context for users and stakeholders. Consequently, it also enables its stakeholder companies and entrepreneurs to exploit its services and tools to improve its competitiveness and strategies (M. Kumar, Graham, Hennelly, & Srai, 2016). The relationship between smart cities and innovative and technological development has been extensively studied by analyzing their impacts on citizens (Gagliardi et al., 2017; Neirotti, De Marco, Cagliano, Mangano, & Scorrano, 2014) and on the economic and social development of the city (Allam & Dhunny, 2019; Leitheiser & Follmann, 2020; Yigitcanlar, Kankanamge, & Vella, 2020).

Companies interact in an innovative and technological context, with ample room for maneuvering and growth, building on the urban potential as a driver and tool for their growth strategies (Neumann et al., 2019). To reach these goals, the smart city concept encompasses factors that include information and communications technology (ICT) and Internet of Things (IoT) elements such as fiber optic networks, sensors and connected devices, and open data analytics (Gagliardi et al., 2017). In addition to these factors, smart cities must also consider the concept of an innovative environment and the relationship between business and urban development (Richard Florida et al., 2017; Leitheiser & Follmann, 2020). Existing smart cities combine skills, knowledge, and capital, providing an innovative technological structure that allows stakeholders to be recombined into new and productive forms (K. Paskaleva & Cooper, 2018; Tachizawa, Alvarez-Gil, & Montes-Sancho, 2015). Furthermore, the current concept of smart cities is expected to facilitate information gathering among citizens, entrepreneurs, and companies. To this end, cities can provide public services and a new competitive edge for stakeholders more efficiently and sustainably (Almirall et al., 2016), in addition to a new urban vision of growth with local and international implications. It assumes that urban innovation is closely related to and interacts with the whole urban ecosystem. Typical smart city projects involve not only large multinationals and local authorities but also local companies and start-ups, who transfer general technological solutions and promote innovation (Caragliu & Del Bo, 2019). The relationship between innovation and cities also features in the studies by Adler et al. (2019), who consider cities at the center of innovation and entrepreneurship processes. He evidences the reflections of Jane Jacobs in her recent urban research on the role of the city and those of Joseph Schumpeter on innovation and entrepreneurship. From this perspective, smart cities are perceived as collaborative ecosystems that facilitate innovation by connecting citizens, governments, businesses, and educational institutions. These innovative clusters foster the development of high-added-value activities and innovative companies (Appio, Lima, & Paroutis, 2019).

Consequently, this relationship impacts the management and development of local companies that can manage the services offered by the city, which is becoming more and more user-centric (Xiahou, Yuan, Xie, Skibniewski, & Li, 2020). The acquisition and perception of these services improve the development and growth opportunities of innovative companies, which consequently impact the city's entire economic and social environment. With entrepreneurship based on local knowledge and supported by the local ecosystem, the role of local companies and knowledge in the smart city transition is still at the center of the debate (Datta, 2015; Kummitha, 2018; Kummitha & Crutzen, 2017, 2019). This debate helps to strengthen the empirical evidence provided in this chapter by highlighting how the implementation of digital technology in cities influences the business and entrepreneurial environment in current smart cities.

Today, smart city projects have the potential to transform traditional urban systems and create an environment that fosters economic growth and innovation. By leveraging technology and data-driven solutions, smart cities can provide more efficient services to citizens, optimize existing infrastructure, and encourage collaboration among different economic actors (Linde et al., 2021; Vanolo, 2014). This advancement leads to a more favorable economic environment for entrepreneurs and businesses of all sizes, particularly those focused on sustainability and social impact. In addition, as we will see in the next paragraph, smart cities affect different perspectives, such as female entrepreneurship, start-ups, and green and innovative companies, by providing opportunities, technologies, and a balanced environment tailored to user needs. These interactions help to create a more diverse and inclusive business community that can generate new ideas and drive economic growth. The impact of smart cities on entrepreneurship has been significant, but there is still room for growth and improvement. For example, smart cities can further support entrepreneurship by providing access to funding, mentorship, and other resources and by developing a synergy that positively influences the perception and development of the city. Moreover, there is a need to ensure that smart city initiatives are inclusive and accessible to all members of the community, particularly those in underserved areas. By doing so, smart cities can continue to catalyze economic development and entrepreneurship while promoting social and environmental sustainability.

6.4 Local Growth and Economic Outcomes

The rise of digital, global, and sustainable cities, and their consequent effects on the national and international environment (Burns, Fast, Levenda, & Miller, 2021; Sassen, 2013), have been accompanied by cutthroat international and intercity competition to boost local growth (Glaeser, 2000; Goldstein & Drucker, 2006; Harvey, 1989; Shelton, Zook, & Wiig, 2015). This change has not only altered the distribution and economic balance of territories but also transformed the way of doing business, opening up new scenarios and opportunities. This chapter discusses the interaction between smart cities and entrepreneurial ecosystems to highlight the economic outcomes of smart city projects. Specifically, building on the Italian context and in accordance with the policies of the European Union (see Methodological Appendix), this chapter sheds light on the effect of smart city implementation on the business and entrepreneurial environment.

The next paragraphs rely on empirical and factual data presented in this book's Appendix to discuss and provide an overview of smart cities' current economic relationships and outcomes. This study analyzes 30 smart city projects from the past decade (2010–2021) to investigate the interaction between smart cities and the economic and business environment, focusing on variations within-country cities. Italy serves as a prominent case study due to the distribution of its smart city projects throughout the country, which ensures heterogeneity in the analysis sample.[1] Furthermore, Italy has been extensively discussed in the smart city transition debate (e.g., Christofi et al., 2021, p. 951; Vanolo, 2014, p. 884), making it a valuable subject of study. Empirical and factual data presented in this book's Appendix are relied

[1]https://www.europarl.europa.eu/RegData/etudes/etudes/join/2014/507480/IPOL-ITRE_ET(2014)507480_EN.pdf

upon to provide an overview of the current economic relationships and outcomes of smart cities. The following paragraphs propose different levels of analysis to explain the economic effects on various economic areas, focusing on the city's economy, green companies, female entrepreneurship, start-ups, and innovative companies.

6.4.1 Economic Environment

Smart cities are transforming how people live and work, and their impact on the economy is increasingly significant, influencing our lives and choices (Vanolo, 2014). By utilizing the latest technologies and data analytics and promoting a citizen-centric environment, smart cities enable businesses to optimize their operations and improve productivity (M. Kumar et al., 2016; Zhang, Zhao, & He, 2020). This, in turn, drives economic growth and creates a vibrant and attractive environment with numerous job opportunities (De Jager, Mthembu, Ngowi, & Chipunza, 2017). In addition, the availability of smart infrastructure and amenities in current cities allows them to attract and promote new businesses, including multinational companies, to set up their headquarters. These companies recognize that a smart city's advanced technology infrastructure can provide them with a competitive advantage in terms of improving their business operations, logistics, and supply chain (Tachizawa et al., 2015).

This scenario positively influences the urban landscape and creates employment opportunities for local residents and individuals from other countries looking for work in a city with a high standard of living and quality of life. This is supported by empirical evidence (refer to Table A3 in the Appendix, Model I), which demonstrates that the progress made in digital implementation and smart city projects has a positive impact on the employment rate of a city. This leads to an economically favorable environment that benefits both citizens (who will beneficiate from a well-established economic environment and have access to employment options) and attracts and retains human capital from other urban areas, contributing to the growth of the local economy (Abid et al., 2022). Thus, it is assumed that the advancement of smart cities contributes to job opportunities within the local ecosystem and positively affects the smart city and entrepreneurial environment. This assumption is supported by recent trends, as employment opportunities have increased significantly in well-implemented smart city projects, resulting in greater economic output for both the cities and the overall economy (Abid et al., 2022, p. 10). Additionally, the impact of smart cities on employment and economic growth is not limited to attracting large corporations and established businesses. It also encourages the creation of small businesses, start-ups, and innovative enterprises that cater to specific user needs, including green and sustainable businesses, youth entrepreneurship, and female entrepreneurship.

6.4.2 Business Creation

The intrinsic nature of smart cities includes various social aspects of the city, such as offering public amenities and facilities to ensure the quality of life, particularly in terms of mobility, education, health, public spaces for interaction, and labor and social integration, which are strongly linked to the concept of smart cities, widely considered in the literature (Evangelista, Guerrieri, & Meliciani, 2014; Maria Lluïsa Marsal-Llacuna, Colomer-Llinàs, & Meléndez-Frigola, 2015). Following the recent academic debate (Kummitha, 2019; Scornavacca, Paolone, Za, & Martiniello, 2020), we know that the development of smart practices in cities may be mostly positive for economic development and the creation of new business ventures, but the opposite degrowth could result if these practices become overdeveloped (Datta & Odendaal, 2019). Since smart city policies do tend to make cities more efficient and socially adaptive, the local entrepreneurial context is also expected to be fostered (Cohen, Almirall, & Chesbrough, 2016), mainly through a general improvement of services and functions and the positive fallout from the relationship between the city and its citizens and users (Caragliu, del Bo, & Nijkamp, 2011). This theoretical state of the art is also confirmed by empirical results that show that smart city advancement is linked to the economic context and drives the creation of new businesses (refer to Table A3 in the Appendix, Model II).

For example, smart city projects in Italy strongly influence business creation by developing an advanced economic environment that drives entrepreneurial activity (Marchesani, Iaia, Masciarelli, & Christofi, 2022). This impact is not limited to economically advanced cities like Milan, Turin, and Bologna. In fact, new hubs with a strong entrepreneurial drive are emerging, such as Florence, Genoa, and Bari. This geographical shift is linked to the development of smart cities, which has first affected large urban centers that are more open to smart city practices, and then spread to less economically developed areas, leading to a new entrepreneurial geography (EU Commission). As a result, cities that embrace smart city initiatives become highly adaptive and attractive to entrepreneurs and the workforce. Fig. 4 highlights the interaction between smart city advancement and business creation in the past decade.

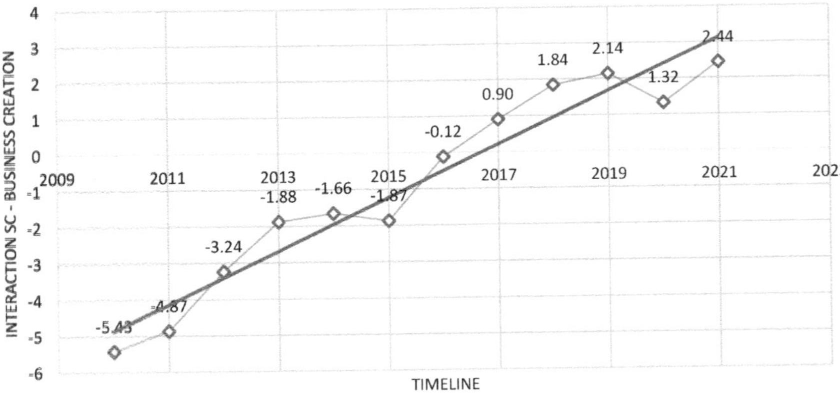

Fig. 4. Interaction Between Smart City Advancement and Business Creation. *Source*: Created by author.

This result contributes to the economic debate on cities' entrepreneurial and economic environments. Furthermore, it adds value to the literature on the intersection of smart cities, entrepreneurship, and human capital by making important theoretical contributions and highlighting the importance of smart city projects in influencing the economic environment (Betz et al., 2016; S. Y. Tan & Taeihagh, 2020; Wiig, 2016).

6.4.3 Start-Ups

The growth of smart cities has had a significant impact on entrepreneurial activity and the development of start-ups, especially those characterized by a youth entrepreneurship component. Start-ups are defined as companies or organizations in their early stages of development and are typically characterized by a focus on innovation and the pursuit of growth (Nanda & Rhodes-Kropf, 2013; Schoof, 2006). As start-ups are often associated with new and emerging technologies, they find fertile ground in the smart city concept, which offers an agile and highly technological environment suitable for the nature of start-ups (Bania, Eberts, & Fogarty, 1993; Chen, Gompers, Kovner, & Lerner, 2010; Nanda & Rhodes-Kropf, 2013). Moreover, in an advanced innovative context, start-ups benefit from advanced infrastructures, such as high-speed internet and 5G networks, which can help them launch and scale their businesses, as well as from a highly advanced entrepreneurial ecosystem that guides entrepreneurship and offers a fertile fabric for entrepreneurial development (Christofi et al., 2021; Marchesani, Masciarelli, & Quang, 2022). Moreover, thanks to their ability to incorporate and attract talent, workforce, and innovation (also due to the active role of universities in the local ecosystem, see Chapter 7), cities are generating a strong impact on the economic scenario by creating new and dynamic businesses (Drach-Zahavy, 2004; Goldstein & Drucker, 2006) that have major economic repercussions on youth entrepreneurship (Marchesani, Masciarelli, & Quang, 2022).

Based on the current academic debate and building on novel empirical investigation, this chapter highlights and confirms that smart city projects are generally associated with a strong relationship toward the creation of start-ups by individuals who move to the city (or who have arrived in the city during their studies) and generate a positive relationship between the city and entrepreneurship (refer to Table A3 in the Appendix, Model III). Furthermore, this positive relationship contributes to the economic environment as entrepreneurship plays a vital role in cities' trajectories concerning economic growth and the perception of the city among its stakeholders. Finally, the results demonstrate how the relationship between digital development in cities and the creation of start-ups constantly grows and diverges from classic urban development (M. E. Porter, 1996), highlighting how the context of the smart city drives entrepreneurship, youth entrepreneurship, and the development of technological start-ups (Marchesani, Masciarelli, & Quang, 2022).

6.4.4 Innovative Companies

The vision of a smart city has been increasingly integrated into urban policies worldwide, influencing the economic environment, business creation, and innovation development in cities (Benevolo, Dameri, & Auria, 2016; Sofronijevic, Milicevic, & Ilic, 2014). Today, the advancement of smart cities worldwide is reshaping the offer of a technological and economic environment tailored to companies and talent, which are considered key elements in orchestrating smart city practices (Linde et al., 2021). Standardizing this environment benefits smart cities by competitively attracting and promoting innovative companies and providing them with a more advanced environment (Adler & Florida, 2021; Richard Florida et al., 2017).

This perspective is supported by empirical evidence highlighting that a highly advanced technological environment impacts the innovative ecosystem, promoting and developing innovative companies that will create a bilateral relationship by utilizing (and contributing to) the urban context (refer to Table A3 in the Appendix, Model IV). Specifically, the city's relationship with innovative companies is implicit in the development of smart cities, which, by offering an optimal environment for them, attracts them to their city, consequently producing a benefit for the city itself. Therefore, developing practices tailored for the creation of an innovative entrepreneurial ecosystem is vital in the dynamics of smart cities (Kummitha, 2018, 2019; Kummitha & Crutzen, 2019). This result provides important insights and adds value to the current literature on both urban innovation (Adler & Florida, 2021; Adler et al., 2019; Richard Florida, 2002b) and smart cities (Kummitha, 2018, 2020; Kummitha & Crutzen, 2017; Marchesani, Masciarelli, & Quang, 2022; Sharifi et al., 2021) and offers significant policy implications for policymakers regarding the success of smart city projects that would benefit stakeholders and the urban system in promoting economic growth and innovation.

6.4.5 Green Companies

Smart cities are transforming the way businesses operate, with a focus on sustainable and green practices to meet the objectives of smart city implementation (Martin, Evans, & Karvonen, 2018; Ooms et al., 2020). The concept of a smart city is centered on the use of advanced technologies to improve quality of life, reduce environmental impact, and enhance economic competitiveness. Therefore, one crucial aspect of smart cities is the integration of sustainable practices into business transformation and development.

This chapter discusses and provides evidence (refer to Table A3 in the Appendix, Model V) that companies located in advanced smart cities are more inclined to adopt sustainable practices such as ISO 9001, ISO 14001, and the EU Eco-Management and Audit Scheme (EMAS) certifications, indicating that the company is actively working to reduce its environmental impact, improve resource efficiency, and promote sustainability (Abid et al., 2022). Smart city practices offer a range of benefits for businesses, including smart energy management systems that reduce energy consumption and carbon footprint, smart transportation systems that enable efficient movement of goods and people, and smart waste management systems that promote recycling and reuse (Chu, Cheng, & Yu, 2021; Khatoun & Zeadally, 2016). Adopting sustainable practices also has economic benefits, as companies embracing sustainability are more attractive to investors and customers who are increasingly interested in sustainability. Sustainable practices can also reduce costs associated with resource consumption and waste management, improving the bottom line and attracting funds from the European Union (Arbolino & Boffardi, 2017; Vanolo, 2014).

Smart cities provide the impetus for businesses to adopt sustainable practices, with the integration of advanced technologies, collaboration and innovation, and economic benefits encouraging companies to develop sustainable and green practices (Leitheiser & Follmann, 2020). This contributes to the overall sustainability of smart cities and promotes a greener, more sustainable future.

6.4.6 Gender Entrepreneurship

Following the existing relationship between smart cities and business creation, and according to the current smart city debate (Neumann et al., 2019; Sofronijevic et al., 2014), we know that the dimensions of a smart city contribute to the urban environment by influencing its development and economic growth (Knudsen et al., 2007). Moreover, this relationship has an even deeper conception founded on the objectives of equity and well-being of the city, which influences and drives female entrepreneurship in the city (Kogut & Mejri, 2022).

Female entrepreneurs play a significant role in socioeconomic development and are directly affected by the urban context (Allen & Bartlett, 2022; Kimbu & Ngoasong, 2016; Rosca, Agarwal, & Brem, 2020). Specifically, in the relationship between the smart city and the economic environment, recent academic research highlights the link between urban livability and female entrepreneurship, identified as the most sensitive actors to the "living dimension" in the city (As introduced in Chapter 3) (Collins & Low, 2010; Nguyen, 2021; Poggesi, Mari, & De Vita, 2016). Moreover, the influence of the smart living dimension also emerges in the urban literature, assuming that women are more willing to advance entrepreneurial activity in a safe and advanced environment (Martínez-Rodríguez, Quintana-Rojo, Gento, & Callejas-Albiñana, 2022). According to this state-of-the-art, this chapter discusses the empirical results (refer to Table A3 in the Appendix, Model VI) to highlight a direct relationship between the advancement of smart city projects and the promotion of female entrepreneurship. The characteristics of the city embedded in the smart city dimensions, such as quality of life, technology, and a well-established economic environment, affect business creation in cities and promote female entrepreneurship, which tends to be more active in a city with a well-developed smart city (Guzman & Kacperczyk, 2019; Kogut & Mejri, 2022). Fig. 5 offers a clear picture of the interaction between smart city implementation and female entrepreneurship. It (i) highlights the distribution of female entrepreneurship across the country and (ii) the interaction

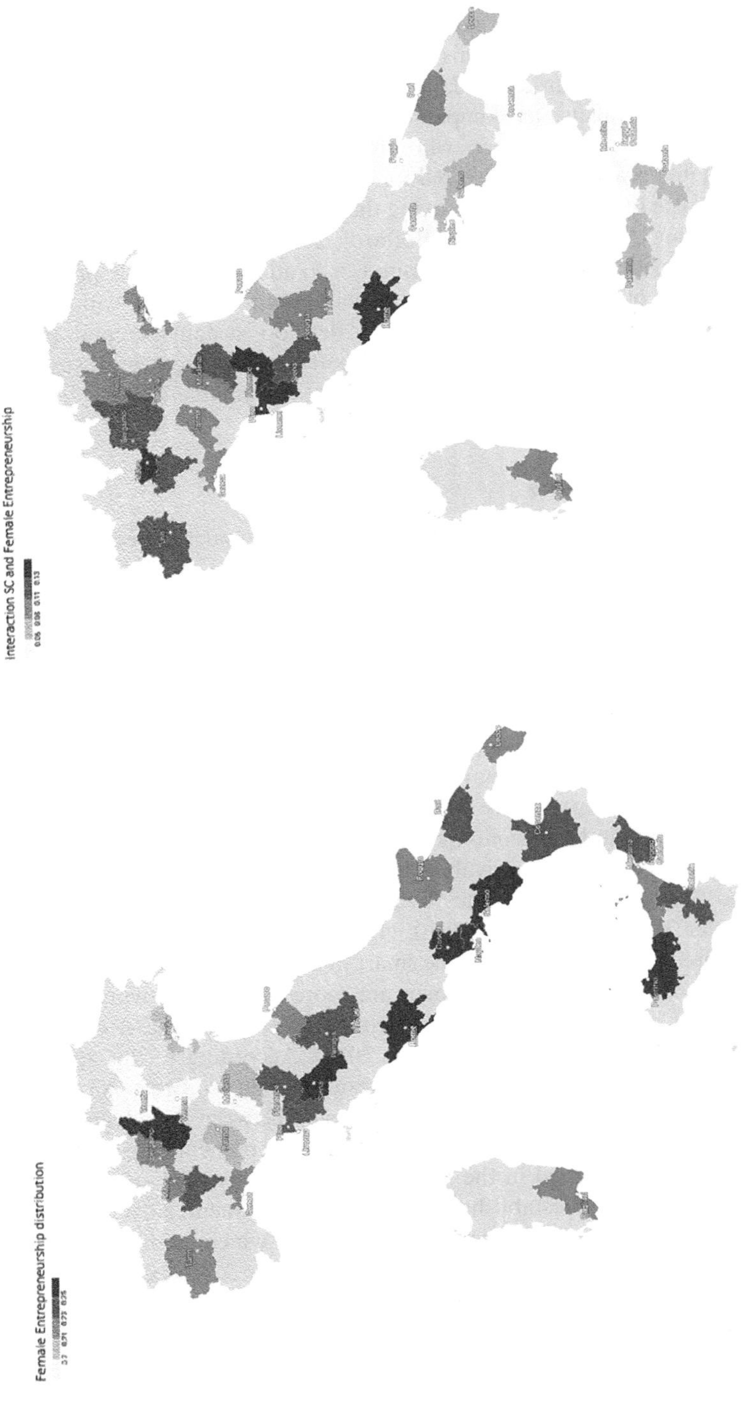

Fig. 5. Interaction Between Smart City Advancement and Female Entrepreneurship. *Source:* Created by author.

between smart cities and female entrepreneurship over the years, showing that female entrepreneurship is becoming localized and growing based on the more favorable urban environment.

This conjecture highlights several practical and theoretical contributions as it shows that smart city advancement (as a context) and the urban and economic environment (as a driver) increase the probability of female entrepreneurship and, more broadly, entrepreneurial activities establishing themselves in cities where they can live and base their life.

6.5 Entrepreneurial Ecosystems Boost Smart City Projects

Entrepreneurial ecosystems and the development of smart cities as technological and data ecosystems (see Chapter 2) are increasingly interconnected, generating positive externalities for both ecosystems that positively impact the current and future development of smart cities (Camboim et al., 2019; Kummitha, 2018; Richard Florida et al., 2017). Entrepreneurial ecosystems are defined as the interconnected elements that contribute to the creation, growth, and sustainability of start-ups and small businesses (Stam & van de Ven, 2021) and comprised of innovative companies, talented individuals, and supportive institutions are becoming increasingly important drivers of smart city transformation (Kummitha & Crutzen, 2019). The entrepreneurial ecosystem directly affects the smart city transformation as it serves as an engine for the smart city projects' technological, managerial, and technical implementation (De Jager et al., 2017; Zhao, Fashola, Olarewaju, & Onwumere, 2021).

The development of an innovative entrepreneurial ecosystem has a significant impact on smart city transformation from different perspectives. Firstly, innovative entrepreneurial ecosystems drive smart city transformation through the development of new technologies and services (Kummitha, 2018). For instance, transportation companies are creating new solutions to reduce congestion and improve mobility, such as ride-sharing platforms and electric vehicles (Benevolo et al., 2016). Likewise, companies in the energy sector are developing innovative systems to generate, store, and distribute renewable energy (Colding & Barthel, 2017). These technologies not only enhance citizens' quality of life but also reduce cities' environmental impact (Khatoun & Zeadally, 2016).

Another key area where entrepreneurial ecosystems contribute to smart city transformation is data collection and analysis (Stock, 2011). These companies are leveraging the vast amounts of data generated by cities to develop new insights and solutions to complex urban challenges. For example, they are using data to optimize traffic flow and reduce energy consumption in buildings (Jindal, Kumar, & Singh, 2020). They are also creating new tools for citizen engagement, such as mobile apps and social media platforms, to enable citizens to participate more fully in the development of their cities (Wiig, 2016).

Finally, an advanced entrepreneurial ecosystem is essential for developing and implementing smart city infrastructure, as it drives innovation, facilitates the integration of new technologies and services, and promotes sustainable and efficient solutions (March & Ribera-Fumaz, 2016). By developing new solutions for the collection, storage, and analysis of data, these companies are enabling cities to become more efficient, sustainable, and resilient. In addition, they are building the networks and systems that underpin smart city initiatives, such as sensor networks for real-time monitoring of air and water quality, and intelligent energy grids for more efficient energy management (Anttila & Jussila, 2018; Asthana & Dwivedi, 2020; Marchesani, Iaia, et al., 2022).

Thus, as previously discussed, the advancement of different dimensions of a smart city affects both the environment and economic growth of a city. At the same time, the presence of a thriving technological entrepreneurial ecosystem brings additional advantages to the city (Abid, Marchesani, Ceci, & Masciarelli, 2023; Kummitha, 2018).

Specifically, a technological entrepreneurial ecosystem provides a competitive advantage to a city as it drives technology innovation and supports the transformation process from within. This vision aligns with the current academic debate (Adler & Florida, 2021; Batabyal & Nijkamp, 2019; Kummitha, 2019; Marchesani, Masciarelli, & Quang, 2022) and contributes to the advancement of the urban landscape and the importance of the connection between data, city, and entrepreneurial ecosystems. Furthermore, cities that are able to provide and manage these ecosystems properly can emerge and lead the transition to smart cities (Kummitha, 2019; Linde, Sjödin, Parida, & Wincent, 2021). Consequently, keeping these resources within their boundaries makes them unavailable to other cities, creating a unique competitive edge (this perspective will be further discussed in Chapter 7) (Taylor Buck & While, 2017).

Thus, in the smart city era, we assume that the relationship between the economic environment and the smart city will influence the trajectory of urban and economic development, and the smart city represents the arena of this future competition (Christofi et al., 2021). As a result, cities must invest in their entrepreneurial ecosystems to create a strong foundation for smart city development (Kummitha & Crutzen, 2017). This includes creating an environment that fosters innovation, attracts and retains talent, and promotes collaboration between businesses and local government (Richard Florida, 2004; Marchesani, Masciarelli, & Quang, 2022). By doing so, cities can create a self-reinforcing cycle that accelerates the growth of both the entrepreneurial ecosystem and the smart city project.

6.6 Conceptual Framework and Final Remarks

This chapter provides an important contribution to the current smart city debate and offers an overview to citizens and policymakers about the current relationship between cities and their economic environments. It summarizes the theoretical and empirical evidence presented in the preceding paragraphs and proposes a conceptual framework. Fig. 6 presents the conceptual framework, which captures the interaction between the smart city and the entrepreneurial ecosystem discussed within this chapter.

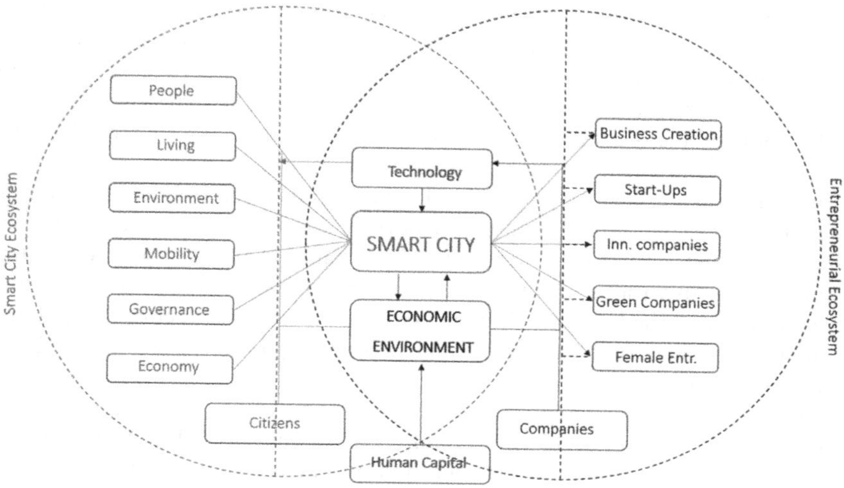

Fig. 6. Conceptual Framework Smart City and Entrepreneurial Ecosystem. *Source*: Created by author.

In the ever-evolving landscape of urban development, the interdependence between the urban environment and the economic ecosystem plays a pivotal role in shaping the future of cities. This intricate and dynamic relationship goes beyond mere coexistence, as it establishes a profound connection that propels cities toward sustainable growth, innovation, and prosperity. The urban environment, with its physical infrastructure, social fabric, and cultural diversity, provides the fertile ground upon which the economic ecosystem thrives. Likewise, the economic ecosystem, encompassing businesses, industries, and entrepreneurial activities, injects vitality into the urban environment, driving economic productivity and shaping its trajectory (Khatoun & Zeadally, 2016; Kummitha, 2018; Kummitha & Crutzen, 2017; Marchesani et al., 2023; Marchesani, Masciarelli, & Doan, 2022).

At the heart of this symbiotic relationship lies the concept of economic agglomeration, which refers to the concentration of economic activities and resources within a specific geographic area. Scholars such as Black and Henderson (1999), Fujita and Thisse (2003), Glaeser (1999), and Naudé, Gries, Wood, & Meintjies (2008) have extensively explored the dynamics of economic agglomeration and its implications for urban development. Traditionally, agglomeration was associated with physical proximity, where cities benefited from the clustering of industries, knowledge spillovers, and economies of scale. However, the advent of smart cities has introduced a paradigm shift in the understanding of economic agglomeration (***Bakici, Almirall, & Wareham, 2013; Betz et al., 2016; Marchesani, Masciarelli, & Doan, 2022).

Smart cities, characterized by their integration of advanced technologies, data-driven decision-making, and sustainable practices, have redefined the notion of economic agglomeration. These cities create an environment that fosters collaboration, innovation, and competitiveness, transcending physical boundaries and embracing the digital realm (Anttila & Jussila, 2018; Camboim et al., 2019; Capdevila & Zarlenga, 2015; Karimikia et al., 2022; Neumann et al., 2019; Scornavacca et al., 2020; Spicer, Goodman, & Olmstead, 2021). The seamless integration of digital infrastructure, IoT devices, and real-time data analytics has propelled smart cities to the forefront of economic transformation. The interconnectedness of various sectors, such as transportation, energy, healthcare, and education, within the smart city framework generates a network effect, amplifying economic activities and catalyzing innovation (Capdevila & Zarlenga, 2015; Carvalho, 2015; Marchesani et al., 2023).

Furthermore, the relationship between smart cities and the economic ecosystem has garnered significant attention in the literature. Building upon the foundations laid by economic agglomeration studies, researchers have delved into the specific mechanisms through which smart cities influence the economic environment. Richard Florida et al. (2017) highlight the role of smart cities in reshaping relationships within a digital, efficient, and competitive context. The utilization of smart technologies and data-driven approaches enables cities to optimize resource allocation, enhance productivity, and attract investment. This, in turn, creates new forms of economic agglomeration, characterized by clusters of businesses and industries that thrive in the digital landscape (Wiig, 2016).

The emergence of these smart city agglomerations has profound implications for policymakers and economic development practitioners. Understanding the dynamics and drivers of these clusters is essential for formulating strategies that foster growth, innovation, and sustainability. Scholars such as Abid et al. (2022), Christofi et al. (2021), and Marchesani (2022), Marchesani and Masciarelli (2023) have emphasized the need to study and harness the potential of smart city agglomerations to maximize economic benefits and create inclusive urban environments. By leveraging the digital infrastructure and collaborative ecosystems within smart cities, policymakers can nurture entrepreneurship, attract talent, and foster knowledge spillovers, creating a virtuous cycle of economic development (Adler & Florida, 2021; Richard Florida, 2004; Holbeche, 2018; Kummitha, 2018; Marchesani et al., 2023; Sharifi et al., 2021). Moreover, the benefits and challenges of smart city projects are deeply intertwined with their relationship with the economic environment. Kummitha (2019) and Kummitha and Crutzen (2019) emphasize that the implementation of smart cities is contingent upon a robust entrepreneurial ecosystem. The presence of start-ups, incubators, and innovation hubs cultivates an environment that nurtures creativity, experimentation, and risk-taking (Adler et al., 2019; Richard Florida et al., 2017; Kummitha, 2020; Marchesani, Masciarelli, & Doan, 2022; M. Tan, 1999). Conversely, the economic environment plays a crucial role in providing the necessary resources, market opportunities, and support systems for smart.

In this intricate interplay between the urban and economic spheres, their inseparable entwining heralds a future where cities transcend their conventional boundaries as mere physical entities, to become dynamic ecosystems pulsating with economic vitality. As smart cities continue to shape the urban landscape, their profound impact on the economic environment stands as a testament to the transformative potential of technology and innovation.

Embracing this inseparable relationship holds the promise of unlocking limitless opportunities, unleashing human creativity, and charting a course toward sustainable and inclusive prosperity. The harmonious synergy between the urban and economic domains serves as the key to unlocking the full potential of smart city projects, propelling us toward a future where cities emerge as exemplars of economic dynamism and livability for generations to come.

Chapter 7

Urban Attractiveness and the Competitive Edge of the Smart City

Abstract

This chapter focuses on the competitive outcomes of present-day smart cities. It explores how cities can leverage economic and internal advancements to gain a competitive edge over other cities, and attract and retain both internal and external users. In the 21st century, cities engage in a global competition to attract citizens and businesses alike. The attractiveness and competitiveness of cities are directly linked to their ability to offer essential services that support growth opportunities, build economic value, and establish a competitive differentiation. Cities with advanced and innovative environments are more likely to maintain leading positions in an increasingly globalized world. Building upon the previous discussions on urban and economic outcomes, this chapter sheds light on the competitive outcomes of smart cities and the implications of global competition and city attractiveness for internal and external users. This chapter begins by examining the role of marketing innovation and internationalization in driving smart city development and how urban ecosystems and digital technologies can enhance city attractiveness. It emphasizes the importance of being a smart destination for both internal and external users and how it can empower and shape interactions with citizens, talent, companies, and tourists. Additionally, this chapter analyzes the geographical differences in smart city strategies and explores the competitive landscape of smart cities within and across nations. Finally, it discusses the limitations of competitive urbanism, along with the challenges and future prospects of smart city development. This analysis is supported by data and published works from international journals. Overall, this chapter aims to comprehensively understand how cities can leverage smart technology and strategies to enhance their national and global positioning.

Keywords: Smart city; competitive cities; urban competitiveness; city attractiveness; economic advancements; global competition; marketing innovation; digital technologies; smart destination; smart tourism

The Global Smart City, 117–135

doi:10.1108/978-1-83797-575-420231007

7.1 Smart City Competition in the Digital Edge

In today's rapidly evolving world, cities are increasingly competing to attract human capital, talent, businesses, and tourists by promoting efficient and highly technologically advanced environments. The global competition among cities has intensified as urban centers strive to become smart cities, leveraging digital technologies and data-driven solutions to improve the quality of life for their residents and enhance their competitiveness on the international stage. In this section, we will explore the central role that digital technologies play in smart city competition, the underlying factors driving this competition, and the various ways cities can leverage these technologies to gain a competitive edge (Christofi, Iaia, Marchesani, & Masciarelli, 2021; Marchesani, Iaia, Masciarelli, & Christofi, 2022; Taylor Buck & While, 2017).

As explored in the initial section of this book, the advent of digital technologies and the availability of vast quantities of data have become indispensable for the functioning of smart cities. These technologies empower cities to monitor, analyze, and optimize urban services and infrastructure, thereby enhancing efficiency, sustainability, and the overall quality of life for their inhabitants. In addition to these advancements, several other factors contribute to the success and competitiveness of smart cities such as urbanization, environmental sustainability, economic competitiveness, globalization, technological advancement, and citizens' expectation:

- *Urbanization*: Contemporary cities require to reshape the urban environment to meet citizens' and stakeholder needs and new urban dynamics and urbanization play a pivotal role in shaping the competition among smart cities. The rapid growth of urban populations necessitates the development of innovative solutions to manage resources efficiently, ensure adequate infrastructure, and provide essential services to meet the escalating demands. Smart cities strive to effectively address the challenges arising from urbanization, such as transportation congestion, housing shortages, and the need for improved healthcare and education systems (Benevolo, Dameri, & Auria, 2016; Linde, Sjödin, Parida, & Wincent, 2021; Shelton, Zook, & Wiig, 2015).
- *Environmental sustainability*: The concept of efficiency and sustainability play a critical factor contributing to environmental sustainability and influencing the competition among smart cities. The pressing concerns related to climate change and the depletion of natural resources have compelled cities to adopt sustainable practices and mitigate their environmental impact. Smart cities leverage digital technologies to monitor energy consumption, optimize waste management, promote renewable energy sources, and enhance overall environmental resilience (Abid, Marchesani, Ceci, Masciarelli, & Ahmad, 2022; Colding & Barthel, 2017).
- *Economic competitiveness*: As discussed in Chapter 6, economic competitiveness significantly shapes the competition among smart cities. These cities aspire to attract investments, foster entrepreneurship, and cultivate innovation-driven ecosystems. They leverage digital infrastructure, data-driven decision-making,

and connectivity to create an environment conducive to business growth and economic prosperity. Furthermore, smart cities focus on nurturing talent, fostering research and development, and facilitating collaborations between academia, industry, and government to drive economic advancement (Kummitha, 2018, 2019; Marchesani, Masciarelli, & Quang, 2022).

- *Globalization*: The increasing interconnectedness of the global economy has heightened the competition among cities to attract investments, businesses, and high-skilled talent. In the context of smart cities, globalization plays a pivotal role as cities strive to position themselves as attractive destinations for international businesses and entrepreneurs. By leveraging digital technologies and data-driven insights, smart cities can create an environment that fosters global collaborations, encourages foreign direct investments, and facilitates international trade. Moreover, being well-connected to global networks enhances a city's reputation and competitiveness on the global stage (Burns, Fast, Levenda, & Miller, 2021; Carvalho, 2015; Christofi et al., 2021, p. 972).
- *Technological advancements*: The rapid pace of technological advancements has provided cities with access to innovative digital solutions that enhance their competitiveness. Smart cities leverage these technological advancements to improve their urban services, infrastructure, and overall functionality. By embracing cutting-edge technologies such as artificial intelligence (AI), Internet of Things (IoT), and big data analytics, cities can optimize resource allocation, enhance mobility and transportation systems, and deliver personalized services to their residents. Technological advancements not only contribute to the efficiency and effectiveness of smart cities but also position them as pioneers in adopting transformative technologies (Kummitha, 2018; Mouton & Burns, 2021; Vial, 2019).
- *Citizen expectations*: With the increasing prevalence of digital technologies in everyday life, citizens have higher expectations for their cities to provide efficient, responsive, and personalized services. Smart cities recognize the importance of meeting citizen expectations and strive to deliver enhanced experiences to their residents. By leveraging digital platforms, data analytics, and citizen engagement tools, smart cities can gather feedback, address concerns, and tailor their services to meet the specific needs of their citizens. Meeting and exceeding citizen expectations not only enhances the quality of life for residents but also contributes to the overall competitiveness and attractiveness of a smart city (Borrás & Edler, 2020; Nam & Pardo, 2011a; Tura & Ojanen, 2022).

To gain a competitive edge in the smart city arena, cities must develop strategies that effectively leverage digital technologies, and data-driven solutions and promote a proper environment for users and stakeholders. These strategies include developing a well-defined smart city vision and roadmap, building an advanced digital infrastructure, encouraging innovation and collaboration, ensuring data privacy and security, promoting digital inclusion, establishing effective governance and regulatory frameworks, and fostering a culture of continuous learning and adaptation (Marchesani, Iaia, et al., 2022).

Specifically, developing a well-defined smart city vision and roadmap is crucial for cities to understand their goals and objectives in becoming smart city. This provides a clear direction and outlines the necessary steps to achieve these objectives. Simultaneously, cities must invest in a robust digital infrastructure comprising high-speed internet connectivity, IoT networks, and data centers (Oztemel & Gursev, 2020). Such infrastructure supports the implementation of smart city solutions, ensuring seamless connectivity and efficient data transmission and storage. In addition, encouraging innovation and collaboration is essential for the advancement of smart cities. Cities can create innovation hubs to foster collaboration among private sector partners, academia, and other stakeholders. This collaboration facilitates knowledge sharing, resource pooling, and the development of cutting-edge smart city solutions (Marchesani, Masciarelli et al., 2022; Qi & Shen, 2019). However, to maintain public trust and comply with regulations, cities must prioritize data privacy and security. Robust mechanisms must be in place to safeguard sensitive information and protect residents' privacy rights. By ensuring data privacy and security, cities can instill confidence in their smart city initiatives.

In this line, digital inclusion is a crucial aspect of smart city development. Cities must ensure that all residents, regardless of their socioeconomic background, have access to digital technologies. Initiatives such as affordable internet access and digital literacy programs promote equality, social cohesion, and economic opportunities for all residents. Moreover, establishing effective governance and regulatory frameworks is necessary for the successful implementation of smart city solutions. Clear policies and regulations ensure that smart city initiatives align with the public interest, promote ethical data practices, and address potential risks and challenges (Abid et al., 2022; Sharifi, Khavarian-Garmsir, & Kummitha, 2021). Thus, smart cities must foster a culture of continuous learning and adaptation. They should actively seek feedback, monitor performance, and learn from both successes and failures. This adaptive approach allows cities to refine their strategies, address emerging challenges, and adapt to evolving technologies and circumstances (Abid, Marchesani, Ceci, & Masciarelli, 2023). Thus, is assumed that by employing these strategies, cities can gain a competitive edge in the smart city arena. They can enhance the quality of life for residents, drive economic growth, and address urban challenges in a sustainable and inclusive manner. With this foundation, the next section will discuss and explore the role of marketing innovation and internationalization in smart city development.

7.2 Marketing Innovation and Internationalization in Smart City Development

In the context of smart cities, international initiatives are strategically formulated across various domains, such as the economy, environment, mobility, sustainability, citizen engagement, and governance. These initiatives are driven by the adoption of IoT and information and communication technologies (ICTs), which transform city infrastructure and services (Christofi et al., 2021; Gagliardi et al.,

2017; Hollands, 2020). Within this framework, international marketing strategies play a vital role in the development of smart cities, evolving over the past decade to support innovation and internationalization processes (Christofi et al., 2021).

Today, smart cities recognize the importance of internationalization and integrate it into various aspects of their operations. By leveraging IoT and ICTs, smart cities transform their infrastructure and services to enhance efficiency, effectiveness, and sustainability. In this line, international marketing strategies have emerged as crucial tools in the development of smart cities. These strategies have evolved over time to adapt to the changing landscape of innovation and internationalization. They enable smart cities to establish a global presence and position themselves on the international stage while effectively promoting their strengths and capabilities to attract and retain investments, businesses, and stakeholders (Christofi et al., 2021; Jungwoo Lee & Lee, 2014; Oztemel & Gursev, 2020).

In doing so, international marketing strategies have played a pivotal role in enabling smart cities to navigate the internationalization process (Christofi et al., 2021, p. 948). They have empowered cities to transcend their local contexts and effectively engage with the global landscape. By adopting international marketing strategies, cities can showcase their unique strengths, innovative solutions, and transformative potential, positioning themselves as attractive destinations for investment, talent, and collaborations (Adler & Florida, 2021; March & Ribera-Fumaz, 2016). Thus, the integration of these strategies in the development of smart cities has been instrumental in supporting innovation, internationalization, and the transformation of small towns into global metropolises. These strategies play a vital role in positioning smart cities on the international stage, attracting investments, and fostering sustainable growth and development.

However, while the concept of adopting smart city designs holds great appeal for many cities, the actual implementation and engagement of key factors such as citizens, entrepreneurs, and companies present significant challenges (Cardullo & Kitchin, 2019; Vanolo, 2014). The role of governance plays a key role in the successful implementation and development of smart cities, particularly with its citizen-oriented perspective (Tura & Ojanen, 2022). By prioritizing the needs of citizens, governance helps bridge the gap between traditional governance strategies and the evolving requirements of the citizens (Neumann, Matt, Hitz-Gamper, Schmidthuber, & Stürmer, 2019).

This citizen-oriented perspective, coupled with the entrepreneurial role in innovation policy, further enhances the awareness and attractiveness of city branding at the international level. Consequently, investments in marketing become essential to effectively implement internationalization strategies for city governance (Marchesani, Iaia, et al., 2022). By leveraging new communication channels and adopting innovative promotional strategies, governance can establish and strengthen relationships and loyalty with stakeholders. This, in turn, opens up new market opportunities and innovative solutions for businesses and entrepreneurs, while attracting fresh human and financial resources on the international landscape (Breslow, 2021; Christofi et al., 2021).

A substantial body of literature underscores the significance of aligning governance with the needs of citizens and emphasizes the importance of stakeholder engagement in smart city development (Albino, Berardi, & Dangelico, 2015; Kummitha & Crutzen, 2019; Marchesani, Iaia, et al., 2022; Neumann et al., 2019). Through the implementation of international marketing strategies, smart cities enhance engagement with all stakeholders, as well as enhance their attractiveness and competitiveness on the international arena.

By strategically embracing international marketing approaches, smart cities adeptly convey their visionary objectives, ambitious goals, and remarkable achievements to key stakeholders. Through effective communication strategies, these cities establish compelling perspectives that ignite collaboration, attract vital investments, and propel them forward on their transformative development journey. This concerted effort toward innovation, sustainability, and inclusive growth positions these cities at the forefront of urban development, driving the evolution of intelligent, vibrant, and cutting-edge urban landscapes. Through their relentless pursuit of efficiency and the rise of global competition, smart cities serve as beacons of technological prowess, both spearheading and following the creation of advanced urban environments that inspire and shape the contemporary cities (Christofi et al., 2021; Marchesani, Iaia, et al., 2022).

7.3 Urban Ecosystem and Digital Technologies Drive City Attractiveness

As anticipated in the previous section, smart cities are increasingly recognizing the pivotal role played by their urban ecosystem and digital technologies in shaping their attractiveness. As globalization continues to reshape the economic landscape, cities are under pressure to leverage technological advancements and foster an environment that supports innovation, entrepreneurship, and sustainable growth. This section explores the interplay between the technological capabilities of businesses located within a city and the technological infrastructure of the city itself, highlighting how these factors contribute to the competitiveness of the urban ecosystem and drive city attractiveness. This interplay is based on continuous interaction between economic and urban environments.

The technological capabilities of businesses located within a city play a crucial role in determining its competitiveness. In today's interconnected world, businesses that possess cutting-edge technologies are better positioned to succeed in global markets. These technologies enable them to enhance productivity, streamline operations, and develop innovative products and services (Huertas, Moreno, & Pascual, 2021; Khan, Woo, Nam, & Chathoth, 2017; Marchesani, 2022). Moreover, the presence of technologically advanced businesses within a city creates a positive spillover effect on the entire urban ecosystem. In this line, an innovative economic environment plays a key role in the smart city transformation (Marchesani, Masciarelli et al., 2022).

Firstly, technologically advanced businesses attract highly skilled workers, researchers, and entrepreneurs, fostering a talent pool that drives further innovation and economic development. The presence of a vibrant and skilled workforce enhances the overall competitiveness of the city and makes it an attractive destination for both businesses and individuals seeking opportunities for growth and collaboration (Kummitha & Crutzen, 2019; Verginer & Riccaboni, 2021). Secondly, technologically advanced businesses often form collaborative networks and clusters within a city. These networks promote knowledge sharing, collaboration, and resource pooling, leading to increased innovation and productivity gains. The synergistic effects of such networks create a fertile ground for the emergence of new ideas, start-ups, and entrepreneurial activities, further enhancing the competitiveness of the urban ecosystem (Richard Florida, Adler, & Mellander, 2017).

In addition to the technological capabilities of businesses and the economic environment, the technological infrastructure of the city itself plays a critical role in driving its attractiveness and competitiveness. A well-developed and integrated technology infrastructure enables the seamless flow of information, efficient transportation, effective communication, and reliable public services, all of which are vital for attracting businesses, residents, and tourists (Marchesani, 2022; Romão, Kourtit, Neuts, & Nijkamp, 2018).

Firstly, robust digital connectivity and broadband infrastructure are essential for businesses to thrive in the digital age. High-speed internet access and reliable communication networks facilitate the exchange of information, enable remote work and collaboration, and support the implementation of smart technologies and IoT applications. Cities that invest in state-of-the-art digital infrastructure create an enabling environment for businesses to leverage advanced technologies and gain a competitive edge in the global market (Angelidou, Politis, Panori, Barkratsas, & Fellnhofer, 2022; Betz, Partridge, & Fallah, 2016; Cimbaljević, Stankov, & Pavluković, 2019).

Secondly, smart city initiatives that leverage digital technologies contribute significantly to the attractiveness of a city. Smart city solutions, such as intelligent transportation systems, energy-efficient buildings, and data-driven governance, enhance the quality of life for residents, promote sustainability, and optimize resource allocation. The integration of digital technologies into urban services and infrastructure improves efficiency, reduces costs, and enhances the overall livability of the city, making it an appealing destination for businesses and individuals alike (Christofi et al., 2021, p. 972).

Today, the innovative economic and urban environment are interdependent and create a mutually reinforcing relationship within the urban ecosystem. Businesses rely on the city's technological infrastructure to operate efficiently and access markets, while the city benefits from the innovative capacity and economic contributions of businesses. This relationship lends to a double perspective as businesses thrive when they have access to reliable and advanced technological infrastructure. This infrastructure enables them to connect with customers, suppliers, and partners, facilitating seamless operations and expanding their market reach. By providing a supportive environment for businesses to leverage technology, cities enhance their competitiveness and position themselves as attractive investment destinations (Adler & Florida, 2021; Bakici, Almirall, & Wareham et al., 2013; Capdevila & Zarlenga, 2015; Cohen, Almirall, & Chesbrough, 2016; Marchesani, Masciarelli et al., 2022; Seyfi & Hall, 2020).

On the other hand, businesses contribute to the technological development of the city by investing in research and development, fostering innovation, and driving the adoption of emerging technologies. Their technological capabilities and expertise enrich the local ecosystem, attract talent, and spur further technological advancements (Bartoloni et al., 2022; Linde et al., 2021; Vinod Kumar, 2020). The resulting ecosystem of collaboration and innovation reinforces the city's attractiveness and competitive advantage.

Thus, it assumes that the symbiotic relationship between the innovative economic and urban environment forms a strong foundation within the urban ecosystem. Businesses depend on the city's technological infrastructure to operate efficiently and expand their market reach, while the city benefits from the economic contributions and innovative capacity of businesses. This dynamic relationship creates a double perspective as cities that provide reliable and advanced technological infrastructure empower businesses to thrive and position themselves as attractive investment destinations. Simultaneously, businesses contribute to the city's technological development by investing in research, fostering innovation, and driving the adoption of emerging technologies. This collaboration and innovation ecosystem further enhances the city's attractiveness and competitive advantage.

7.4 Smart Destination and City Attractiveness

This section explores the notion of smart destinations and their contribution to the overall attractiveness of cities (Boes, Buhalis, & Inversini, 2015; Jovicic, 2019; D. Wang, Li, & Li, 2013). Building upon the symbiotic relationship between innovative economic environments and urban settings discussed in the previous chapter, this section focuses on the role of smart destinations in attracting talents, citizens, tourists, and businesses. By analyzing the dynamic interplay between technological infrastructure, collaboration, and innovation ecosystems, we can gain insights into how cities can position themselves as compelling investment destinations.

Smart destinations are characterized by their effective integration of advanced technologies, data-driven decision-making processes, and innovative solutions aimed at improving the quality of life for residents and visitors (Gretzel, Koo, & Gretzel, 2021; Huertas et al., 2021; J. Wang, Xie, Huang, & Morrison, 2020). The concept of smart destinations revolves around the idea of leveraging advanced technologies, data-driven decision-making, and innovative solutions to create environments that enhance the quality of life for residents and visitors.

These destinations integrate smart city initiatives into their urban fabric, fostering economic growth and development. The presence of reliable and advanced technological infrastructure plays a crucial role in attracting businesses, enabling them to operate efficiently and expand their market reach. Furthermore, businesses contribute to the city's technological development by investing in research, fostering innovation, and driving the adoption of emerging technologies (Dameri & Rosenthal-Sabroux, 2014; Sharifi et al., 2021). This collaboration and innovation ecosystem further enhances the city's attractiveness and competitive advantage.

One aspect of city attractiveness is the ability to attract and retain talented individuals. Smart destinations, with their focus on technology, innovation, and sustainability, are particularly appealing to a new generation of skilled professionals seeking dynamic and progressive environments (Dotti, Fratesi, Lenzi, & Percoco, 2014; Sinkiene & Kromolcas, 2010). These cities invest in research and development, foster innovation through collaborative networks, and provide opportunities for personal and professional growth. By nurturing a culture of innovation and attracting a diverse pool of talent, smart destinations fuel further economic and social development.

Moreover, smart destinations prioritize citizen-centric approaches to urban planning and governance. By leveraging technology, these cities empower residents to actively participate in decision-making processes and contribute to the development of their communities. Through digital platforms and smart services, citizens can access information, provide feedback, and engage with local authorities. This participatory approach fosters a sense of ownership and belonging, leading to increased citizen satisfaction and a higher quality of life. By prioritizing citizen engagement, smart destinations create environments where residents can thrive and actively contribute to the city's growth (Kummitha & Crutzen, 2019; Saxenian, 2002; Verginer & Riccaboni, 2021).

Thus, it assumes that smart destinations play a vital role in enhancing the attractiveness of cities by leveraging advanced technologies, collaboration, and innovation ecosystems. The presence of reliable and advanced technological infrastructure empowers businesses to thrive, positioning cities as attractive investment destinations. Additionally, the focus on talent attraction, citizen engagement, and community development further contributes to a city's overall attractiveness. The following subsections will explore in more detail the specific impacts of smart destinations on talents, citizens, tourists, and businesses, and how cities can capitalize on these factors to become vibrant and prosperous urban centers.

7.4.1 Citizens

In recent years, there has been a notable shift in urban dynamics, with an increasing number of individuals gravitating toward smart cities. The allure of these technologically advanced urban centers lies in their ability to offer a higher quality of life and improved opportunities for residents. The reasons that drive citizens to seek out smart cities are multifaceted (de Falco, Angelidou, & Addie, 2019; Greenwood & Sweetland, 1972; Rosenthal & Strange, 2003).

One key factor is the pursuit of enhanced connectivity and convenience. Smart cities leverage advanced technologies to streamline various aspects of daily life, ranging from efficient transportation systems to digital services that simplify administrative processes (Albino et al., 2015; Benevolo et al., 2016). Citizens are drawn to the seamless integration of these technologies, which ultimately saves them time and effort. Additionally, the availability of digital platforms and apps that provide instant access to information and services empowers citizens to actively engage with their surroundings (Linde et al., 2021).

Another crucial motivation is the desire for sustainability and environmental consciousness. Smart cities embrace eco-friendly practices and employ innovative solutions to address pressing environmental challenges (Abid et al., 2022). From efficient energy management to waste reduction initiatives, these cities prioritize sustainability, resonating with citizens who are increasingly aware of the need for a greener future. By fostering a culture of environmental responsibility, smart cities offer a compelling alternative to traditional urban settings (Angelidou, Politis, Panori, & Barkratsas et al., 2022).

As a consequence, the benefits of living in a smart city are manifold. Improved infrastructure, such as reliable public transportation and smart grids, enhances the overall efficiency and functionality of the urban environment. This, in turn, leads to a higher standard of living and greater convenience for citizens. The availability of diverse job opportunities, particularly in emerging sectors such as technology and innovation, attracts talented individuals seeking career growth and professional development. Additionally, smart cities often boast vibrant cultural scenes, ample recreational spaces, and a sense of community that further contribute to their appeal (Batabyal & Nijkamp, 2019; Bibri & Krogstie, 2020; Mergel, Edelmann, & Haug, 2019; Nesti, 2019; Richard Florida, 2002b).

However, it is essential to acknowledge the potential downsides of the increasing concentration of citizens in smart cities. As urban centers become more attractive, rural areas and smaller cities may face depopulation and economic decline (Jacobs, 1969; Morinière, 2012; Spicer, Goodman, & Olmstead, 2021). This phenomenon creates an imbalance, placing additional strain on resources and infrastructure in already densely populated areas. Furthermore, the rapid pace of technological advancements can lead to concerns about privacy, data security, and the potential for exclusion of certain segments of the population who may not have access to or be proficient in using advanced technologies.

The competition among cities to attract and retain citizens has become increasingly fierce. Municipalities worldwide are vying for residents, aiming to establish themselves as premier destinations for living, working, and enjoying a high quality of life. This competition has prompted cities to invest in innovative initiatives, infrastructure upgrades, and urban development projects (Christofi et al., 2021; Sinkiene & Kromolcas, 2010). By focusing on citizen-centric approaches, smart cities have an edge in this race, as they offer residents an unparalleled range of opportunities and amenities. As a consequence, the impact of attracting and retaining citizens extends beyond the social fabric of a city; it also has significant economic implications. A thriving population fuels economic growth by attracting businesses, generating employment opportunities, and stimulating innovation and entrepreneurship. Moreover, citizen engagement and active participation in civic matters contribute to the overall development and sustainability of the urban ecosystem (Caragliu & Del Bo, 2019; Vanolo, 2014).

In the words of renowned urbanist Jane Jacobs (1961), "Cities have the capability of providing something for everybody, only because, and only when, they are created by everybody." This quote encapsulates the essence of citizen-centric smart cities. By prioritizing the needs and aspirations of residents, smart cities have the potential to create inclusive, livable, and vibrant urban environments that harness the collective power of their citizens.

7.4.2 Talents

As for citizens, the interaction between smart cities and talent has become the focal point of academic debate due to its multifaceted impact. Smart cities possess intrinsic characteristics that, when combined with cultural and recreational amenities, vibrant arts scenes, and diverse communities, enrich the urban experience and make them highly desirable places to live, work, and explore. Specifically, talents are drawn to smart cities due to the myriad of opportunities they offer for personal and professional growth (Adler & Florida, 2021; Kummitha & Crutzen, 2019; Marchesani, Masciarelli et al., 2022; Schultz, 1961).

The advantages for cities in attracting and retaining talent are manifold. First and foremost, talents act as catalysts for economic growth and development. Their skills, expertise, and entrepreneurial spirit drive innovation, job creation, and the growth of key sectors such as technology, healthcare, and finance. Moreover, talents contribute to the diversity and cultural vibrancy of cities, fostering social interactions and creating a multicultural environment (R. Florida & Kenney, 1992; Richard Florida, 2002b; Zhao, Fashola, Olarewaju, & Onwumere, 2021). Additionally, the presence of talents attracts investments from businesses and organizations seeking access to a skilled workforce and a favorable innovation ecosystem (Kummitha, 2018; Marchesani, Masciarelli et al., 2022). This, in turn, stimulates further economic activity and creates a positive feedback loop for the overall development of the city. However, balancing the urban environment with the economic environment poses significant challenges for cities as rapid urbanization and increasing resource demands strain infrastructural capacities, social cohesion, and environmental sustainability.

The current global landscape is characterized by fierce competition among cities to attract and retain talents. Recognizing the transformative power of talents in driving economic growth and prosperity, cities are implementing strategies and policies to position themselves as talent hubs. They invest in creating favorable business environments, enhancing educational and research facilities, and fostering collaboration among academia, the economic sector, and the government (de Falco et al., 2019; D. G. De Silva & McComb, 2012; Pittaway & Montazemi, 2020). Moreover, cities focus on cultivating strong cultural identities, promoting diversity and inclusivity, and offering a high quality of life. By differentiating themselves in these aspects, cities can enhance their attractiveness and gain a competitive edge in talent attraction (Richard Florida, 2002b).

The importance of talents in the development of smart cities cannot be overstated. Talents are the lifeblood of innovation, the driving force behind technological advancements, and catalysts for societal progress (Kummitha & Crutzen, 2019; Marchesani, Masciarelli et al., 2022). Their creativity, expertise, and entrepreneurial mindset are crucial in tackling complex urban challenges and shaping the future of cities. Renowned urbanist Richard Florida (2012) once stated that "the most fundamental advantage cities confer on their inhabitants is the ability to connect and learn from one another. This is the ultimate source of innovation and productivity." Smart cities, by fostering an environment that attracts and nurtures talents, harness this fundamental advantage and pave the way for sustainable growth and prosperity.

Therefore, the rise of smart cities has reshaped the landscape of talent attraction and retention. The convergence of advanced technological infrastructure, innovation ecosystems, and the interplay between the urban and economic environment makes these cities highly appealing to talents worldwide. By capitalizing on recent trends, offering ample opportunities for personal and professional growth, and effectively addressing the challenges of urban development, smart cities can establish themselves as thriving talent hubs. Talents, in turn, play a pivotal role in driving economic growth, fostering innovation, and shaping the future of smart cities (Batabyal & Nijkamp, 2019; Faggian & McCann, 2009; Kummitha & Crutzen, 2019; Marchesani, Masciarelli et al., 2022). As cities vie to become magnets for talents, they lay the foundation for a prosperous and sustainable urban future.

7.4.3 Companies

Despite the interaction between various internal actors such as citizens and talents, the urban ecosystem plays a pivotal role in catalyzing businesses and entrepreneurial activities by providing a favorable environment that fosters innovation, growth, and sustainability. The integration of advanced technologies, data-driven decision-making processes, and collaborative ecosystems renders smart cities highly appealing to businesses seeking to establish and expand their operations (Abid et al., 2023; Bakici, Almirall, & Wareham, 2013; Capdevila & Zarlenga, 2015; Cohen et al., 2016).

One of the primary factors that attract businesses to smart cities is the presence of a supportive and advanced technological infrastructure (Camboim, Zawislak, & Pufal, 2019). These cities boast a robust digital framework that enables seamless connectivity, efficient communication, and real-time data access. This infrastructure empowers businesses to streamline their operations, optimize resource utilization, and enhance decision-making processes. Moreover, smart cities offer a wide array of digital services and platforms that facilitate business transactions, improve customer experiences, and extend market reach (Appio, Lima, & Paroutis, 2019; March & Ribera-Fumaz, 2016; Taylor Buck & While, 2017). The availability of such technological advancements confers a competitive advantage to businesses and incentivizes their establishment within smart cities (Christofi et al., 2021).

Furthermore, the presence of vibrant innovation ecosystems in smart cities is another allure for businesses. These cities foster collaboration between businesses, academia, research institutions, and the government, creating an environment that promotes knowledge exchange, research and development, and entrepreneurial endeavors (Kummitha, 2018, 2019; Marchesani, Masciarelli et al., 2022). Within these ecosystems, valuable networking opportunities, access to talent, and support for start-ups and small businesses abound. By creating a conducive environment for innovation, smart cities effectively attract businesses that seek to tap into cutting-edge research and entrepreneurial talent, ultimately driving growth and economic vitality (Camboim et al., 2019; S. Y. Tan & Taeihagh, 2020).

The advantages for cities in cultivating an environment tailored to business and entrepreneurial activities are substantial, as discussed in the previous chapter. Firstly, businesses contribute significantly to the economic development of cities by generating employment opportunities, stimulating investment, and fostering overall economic growth. The establishment of businesses in smart cities leads to job creation, thereby enhancing the standard of living for residents. Moreover, businesses contribute to the city's tax base, providing revenue that can be invested in infrastructure development, public services, and community initiatives (Abid et al., 2022). Additionally, the presence of businesses in smart cities bolsters the city's reputation as a thriving economic hub, attracting further investments and fostering a positive business climate (Neumann et al., 2019).

In the current global landscape, cities face fierce competition to attract and retain businesses. Recognizing the pivotal role of businesses in driving economic prosperity, cities are implementing strategies and policies to position themselves as attractive business destinations. They prioritize the creation of business-friendly environments by simplifying administrative procedures, offering incentives, and providing support services that foster entrepreneurship and business growth. Furthermore, cities invest in infrastructure development, transportation networks, and smart logistics systems to enhance connectivity and facilitate the movement of goods and services. By cultivating an environment that nurtures businesses, smart cities gain a competitive edge in attracting both local and international enterprises (Albino et al., 2015; Breslow, 2021; Oztemel & Gursev, 2020; Sofronijevic, Milicevic, & Ilic, 2014; Vanolo, 2014).

In conclusion, the interaction between smart cities and businesses is vital for fostering innovation, economic growth, and sustainability. The advanced technological infrastructure and vibrant innovation ecosystems offered by smart cities attract businesses seeking a conducive environment for growth and expansion.

7.4.4 Tourists

The emergence of smart tourism and the evolving dynamics between cities and tourism have significantly reshaped the landscape of travel and leisure in cities. Smart cities have become increasingly appealing to tourists, offering a unique blend of technological advancements, sustainable practices, and enhanced visitor experiences (Gretzel et al., 2021; Marchesani, 2022; Schaefer & Hetman, 2019; Shafiee, Rajabzadeh Ghatari, Hasanzadeh, & Jahanyan, 2021). This section explores the new trends in smart tourism, the relationship between cities and tourism, and how this is redefining the geography of tourism (Buhalis, 2020). Additionally, it delves into the competition among cities to attract tourists, showcasing diverse strategies based on sustainability, technology, and experiential offerings.

One of the key factors driving tourists toward smart cities is the concept of smart tourism. This entails leveraging advanced technologies, data-driven insights, and personalized services to enhance the overall visitor experience. Smart cities integrate these elements into their tourism infrastructure, offering interactive maps, mobile applications, and smart transportation systems that provide real-time information and guidance (Buhalis, 2020; Buhalis & Amaranggana, 2015; Buhalis, O'Connor, & Leung, 2023). This seamless integration of technology enables tourists to navigate the city effortlessly, access tailored recommendations, and engage in immersive experiences. Moreover, smart cities prioritize sustainability, implementing eco-friendly practices, such as renewable energy solutions, waste management initiatives, and sustainable transportation options, all of which contribute to a greener and more responsible tourism experience (Marchesani, 2022).

The relationship between cities and tourism has evolved from a traditional model of visitor attractions to a more integrated and symbiotic partnership. Smart cities recognize the significant economic impact of tourism and actively seek to enhance their attractiveness as tourist destinations (Gafter & Tchetchik, 2017; Krabokoukis & Polyzos, 2020). They understand that tourists not only contribute to the local economy through spending but also play a crucial role in shaping the city's identity and reputation (Romão et al., 2018). As a result, cities are investing in infrastructure development, cultural preservation, and the enhancement of visitor amenities to cater to the diverse needs and preferences of tourists. Additionally, cities are actively engaging with tourists, leveraging digital platforms and social media to facilitate communication, gather feedback, and cocreate experiences (Marchesani, 2022, p. 3). This collaborative approach ensures that tourists feel valued and have a voice in the ongoing development of the city.

The emergence of smart tourism and the increasing attractiveness of smart cities have redefined the geography of tourism. Traditionally popular tourist destinations have faced challenges in meeting the evolving demands of modern travelers, while emerging smart cities have seized the opportunity to position themselves as attractive alternatives (Bastidas-Manzano, Sánchez-Fernández, & Casado-Aranda, 2021; Khan et al., 2017; Nilsson, 2019). This shift in the tourism landscape has led to a dispersion of tourism flows, with tourists seeking out unique and authentic experiences in lesser known destinations that offer a blend of technology, sustainability, and cultural immersion. As a result, smart cities, with their innovative approaches, are gradually transforming the tourism map, promoting a more diversified and distributed tourism ecosystem (Gretzel, Sigala, Xiang, & Koo, 2015).

The competition among cities to acquire and retain tourists has intensified in recent years. Smart cities are employing various strategies to differentiate themselves and attract visitors. Sustainability has emerged as a prominent theme, with cities focusing on eco-friendly practices, green spaces, and sustainable transportation options. By positioning themselves as environmentally responsible destinations, smart cities appeal to conscientious travelers who seek sustainable and low-impact experiences (Nilsson, 2019). Technology is another key driver of competitiveness, as cities invest in smart infrastructure, augmented reality applications, and immersive technologies to create unique and interactive experiences for tourists. From smart museums and digital art installations to virtual reality city tours, technology is leveraged to provide engaging and memorable encounters (Buhalis, 2020). Furthermore, cities are focusing on curating experiential tourism offerings, showcasing their cultural heritage, culinary delights, and local traditions to provide authentic and immersive experiences for visitors.

Thus, it assumes that smart cities have become highly attractive to tourists due to the advent of smart tourism and the evolving dynamics between cities and tourism. The integration of advanced technologies, sustainable practices, and enhanced visitor experiences distinguishes smart cities as desirable destinations (Buhalis & Wagner, 2013; Leung, Law, van Hoof, & Buhalis, 2013; Marchesani, 2022). The competition among cities to acquire and retain tourists has prompted the development of diverse strategies based on sustainability, technology, and experiential offerings.

7.5 Limits of Competitive Urbanism and Smart City Development

As we have seen in the previous sections, the rapid rise of smart cities has captured global attention as urban areas embrace technology-driven solutions to address societal challenges affecting urban competitiveness and attractiveness. While smart city development offers evident benefits, it is crucial to acknowledge and critically examine the limitations and potential drawbacks associated with this approach. This section explores the criticisms and challenges surrounding smart city development, highlighting the importance of balancing the benefits with the potential problems.

Smart city development is not without its challenges and criticisms. Privacy and data security concerns arise as cities become more connected and reliant on data-driven technologies, raising questions about unauthorized access, surveillance, and data misuse (Löfgren & Webster, 2020). Additionally, despite the acknowledgment that smart city initiatives, through the implementation of digital and technological processes, have the potential to widen socioeconomic disparities, this advancement could leave disadvantaged communities behind and reinforce existing inequalities (Wiig, 2016). Technological reliance introduces vulnerabilities, making cities susceptible to cyberattacks, system failures, and obsolescence. Ethical considerations and algorithmic bias also come into play as smart city technologies rely on algorithms that can perpetuate biases and compromise fairness (Kiss & Muha, 2018; Z. Li & Liao, 2018). Finally, sustainability and environmental impact pose challenges as the deployment of new technologies can contribute to increased energy consumption and e-waste generation, demanding a careful balance between technological advancements and genuine sustainability practices (Abid et al., 2022).

One of the primary criticisms of smart city development revolves around privacy and data security. As cities become more connected and data-driven, the collection and utilization of vast amounts of personal data raise concerns about the potential for unauthorized access, surveillance, and misuse. The extensive monitoring of individuals' activities and the integration of various data sources pose significant challenges in protecting citizens' privacy and ensuring the security of sensitive information (Ng & Wakenshaw, 2017).

Moreover, since smart city initiatives have the potential to exacerbate existing socioeconomic disparities within urban areas, the implementation of advanced technologies and infrastructure may primarily benefit affluent neighborhoods or business districts, leaving disadvantaged communities behind. The digital divide, unequal access to resources, and a lack of technological literacy can further marginalize vulnerable populations, creating a digital divide that reinforces social inequalities (Datta, 2015; Datta & Odendaal, 2019).

Additionally, as cities become increasingly reliant on technology, they also become more susceptible to vulnerabilities and disruptions. Reliance on complex technological systems makes cities susceptible to cyberattacks, system failures, or technological obsolescence (Hollands, 2008, 2015; Pittaway & Montazemi, 2020). Moreover, the concentration of critical functions and services in smart city infrastructures increases the potential impact of these disruptions, which can have far-reaching consequences for urban resilience and the well-being of citizens. In this line, smart city technologies often rely on algorithms and AI systems to process data and make decisions. Issues of algorithmic bias and lack of transparency in decision-making processes raise ethical concerns and can lead to unintended consequences, reinforcing existing social inequalities and compromising fairness and justice (Uşaklı, Koç, & Sönmez, 2017).

Finally, while smart city initiatives often emphasize sustainability, there are challenges in achieving genuine environmental benefits. The deployment of new technologies and infrastructure as well as a large number of users and stakeholders placed in cities due to the advance of urban attractiveness can contribute to increased energy consumption, e-waste generation, and resource extraction (Arabindoo, 2020; Linde et al., 2021). Additionally, the reliance on data centers and communication networks can strain energy resources and exacerbate environmental impacts. Smart city development must carefully balance technological advancements with genuine sustainability practices to mitigate potential negative environmental consequences.

Thus, it assumes that balancing the benefits and challenges of competitive urbanism and smart city development is crucial for creating inclusive, resilient, and sustainable cities. Acknowledging the limitations and potential problems allows policymakers, researchers, and urban planners to address the criticisms and develop strategies to mitigate the associated risks. Further research is needed to understand the complex dynamics and long-term impacts of smart city initiatives fully. As urban areas continue to evolve and transform, it is essential to critically evaluate the trade-offs and ensure that technological advancements align with the principles of social equity, privacy protection, environmental sustainability, and ethical decision-making (Datta, 2015; Datta & Odendaal, 2019; Grossi & Pianezzi, 2017; Hollands, 2020; Marchesani, 2022). Only through careful analysis and reflection can cities harness the true potential of smart city development while minimizing the risks and maximizing the benefits for all citizens.

7.6 Challenge and Future Prospectives

As seen in the previous section, the development of smart cities presents a wide array of benefits that can improve the quality of life for citizens, optimize urban infrastructure, and promote sustainable growth. However, it is crucial to strike a balance between these advantages and the challenges associated with implementing smart city initiatives. This entails acknowledging potential limitations and problems and adopting a proactive approach to addressing them. This section will delve into the key aspects that can help navigate these challenges and explore future perspectives in smart cities. Specifically, as cities undergo rapid transformation, it is essential to remain vigilant and attentive to the implications of these changes. This involves focusing on the following aspects:

- *Analyzing the long-term societal and economic impacts of smart city projects*: To ensure the sustainability and resilience of smart cities, it is essential to assess the long-term consequences of their initiatives on society and the economy. This includes considering the potential displacement of jobs due to automation, the impact on social equity and inclusion, and the implications for economic growth and competitiveness (Camboim et al., 2019; Kummitha, 2019).
- *Investigating the effectiveness of different governance models and policy frameworks*: Smart cities require effective governance structures and policy frameworks that can accommodate the complex interplay between technology, urban infrastructure, and the needs of citizens. By examining various models and frameworks, policymakers can better understand which approaches are most suited to their specific contexts and goals (de Falco et al., 2019; Gil-Garcia, Dawes, & Pardo, 2018; Nam & Pardo, 2014).
- *Examining the role of citizen engagement and participation in shaping smart city development*: The success of smart city initiatives depends on the active involvement of citizens in decision-making processes, the cocreation of solutions, and the monitoring and evaluation of outcomes. Engaging citizens in these efforts fosters a sense of ownership and ensures that smart city projects are responsive to the needs and preferences of the people they are meant to serve (Cardullo & Kitchin, 2019; Zavattaro & Sementelli, 2014).
- *Exploring the potential of emerging technologies, such as blockchain and AI, in enhancing smart city capabilities*: As new technologies continue to emerge, it is vital to assess their potential applications in smart city contexts. By exploring

the capabilities of technologies like blockchain and AI, cities can harness their potential to improve urban systems, increase efficiency, and promote innovation.

- *Assessing the scalability and replicability of successful smart city models across diverse geographic and cultural contexts*: The experiences of pioneering smart cities offer valuable lessons for other cities looking to embark on similar journeys. By analyzing the factors that contribute to their success and identifying the challenges they have encountered, cities can develop strategies for scaling and replicating these models in their unique contexts (Christofi et al., 2021; Wiig, 2016).

The dynamic nature of smart city development necessitates a continuous commitment to evaluation, adaptation, and improvement. This involves several initiatives for the cities' management such as (i) continuously monitoring and assessing smart city initiatives, (ii) promoting adaptive management strategies to address challenges and improve outcomes, (iii) building on data-driven decision-making and performance metrics in smart city governance, and (iv) fostering partnerships between academia, industry, and government to drive smart city development.

To achieve these goals, smart cities must regularly monitor and assess their projects, identify areas for improvement, measure progress toward established goals, and use this information to inform decision-making. Additionally, they should effectively address challenges, learn from experience, and continuously refine their strategies to achieve better outcomes. Moreover, knowledge-sharing among cities can facilitate the exchange of best practices, lessons learned, and innovative ideas. This accelerates learning and innovation in the smart city domain, promotes interaction and engagement with various actors, and enables the sharing of expertise, resources, and insights. Ultimately, it fosters a collaborative approach to problem-solving and innovation.

However, despite the fundamental importance of continuous evaluation, adaptation, and improvement, aligning technological advancements with the key principles of smart city development has become paramount (Caragliu & Del Bo, 2019; Caragliu, del Bo, & Nijkamp, 2011; Linde et al., 2021; Vanolo, 2014). In order to fully benefit from smart city development, it is crucial to ensure that technological advancements align with key principles such as social equity, privacy protection, environmental sustainability, and ethical decision-making (Kummitha, 2019; Kummitha & Crutzen, 2017; Pittaway & Montazemi, 2020). This underscores the importance of these principles in smart city development, as it highlights the need to strike a balance between technological advancement and the well-being of all citizens. By placing the needs and interests of citizens at the center of smart city initiatives, cities can ensure that technological advancements contribute to the creation of urban environments that are more inclusive, equitable, and sustainable (Ellison, Bannister, Lee, & Haleem, 2021).

In conclusion, the transformative power of smart city development holds immense potential to reshape urban landscapes and improve the lives of citizens.

To maximize this potential, it is essential to minimize risks and maximize benefits through continuous evaluation, adaptation, and improvement. By fostering collaboration, innovation, and a commitment to inclusive, resilient, and sustainable smart cities, stakeholders can chart a course toward a bright and promising future for urban living. As we move forward, it is crucial for all stakeholders – from government institutions and industry leaders to academia and citizens themselves – to work together in achieving the shared vision of smart cities. This collaborative approach will not only help to address the challenges and limitations that may arise along the way but also ensure that the benefits of smart city development are realized by all members of society.

By placing an emphasis on continuous evaluation, adaptation, and improvement, we can create a future where smart cities are not mere showcases of technological advancements, but rather thriving urban ecosystems that prioritize social equity, environmental sustainability, and the well-being of all citizens. This long-term, holistic perspective on smart city development will be instrumental in maximizing the transformative power of smart cities and paving the way for a more resilient and sustainable urban future. In an era of rapid urbanization and technological progress, the potential of smart cities to revolutionize the way we live, work, and interact with our environment is both exciting and complex at the same time. As we continue to explore and embrace these possibilities, let us not lose sight of the importance of considering humans at the center of the urban environment and collaborating across sectors to create inclusive, resilient, and sustainable smart cities that truly serve the needs and aspirations of all citizens.

Chapter 8

Navigating the Smart Cities: Conclusions and Final Remarks

Abstract

This chapter concludes the book by addressing the complexities and challenges associated with smart cities. Despite the widespread enthusiasm and progressive policies surrounding smart cities, navigating this world remains a desirable yet complex challenge. Drawing on existing literature, this final chapter serves as a synthesis of the key points covered throughout the book and offers practical guidelines for navigating the landscape of smart cities. This chapter provides valuable tips and insights for both citizens and policymakers, recognizing their pivotal roles in shaping the future of smart cities. Additionally, it delves into the current challenges faced by smart city initiatives and discusses potential future directions. Emphasizing the need for ongoing adaptation and innovation, this chapter highlights the importance of overcoming obstacles and capitalizing on opportunities in the rapidly evolving digital age. By offering a comprehensive overview and practical recommendations, this chapter aims to equip readers with the knowledge and tools to navigate the complexities of smart cities effectively. It concludes the book on a forward-looking note, emphasizing the dynamic nature of smart city development and the continuous need for adaptation and innovation.

Keywords: Smart city; final remarks; navigating urban environment; complexity; challenges; guidelines; digital city

8.1 Recap and Key Takeaways

The chapters in this book have explored the internal implementation of smart city projects and the external outcomes with the aim to provide readers with a comprehensive and in-depth overview of the main concepts, outcomes, and case studies related to smart cities. Throughout this book, we have explored the multifaceted aspects of smart city initiatives, examining their benefits and challenges within the

The Global Smart City, 137–159

Published under exclusive licence by Emerald Publishing Limited
doi:10.1108/978-1-83797-575-420231008

context of technological advancements, governance models, citizen engagement, and sustainability. By delving into the technological advancements that shape smart cities, we have uncovered the transformative potential of emerging technologies such as artificial intelligence (AI), the Internet of Things (IoT), and blockchain. These technologies offer innovative solutions to enhance urban services, optimize resource allocation, and improve the overall quality of life for citizens. However, we have also critically examined the risks associated with technological dependency, data privacy, and cybersecurity, emphasizing the need for responsible and ethical practices in the implementation of smart city projects.

In the exploration of governance models, this book has highlighted the importance of collaborative approaches that bring together government entities, private sector organizations, academia, and citizens. Through internal implementation and data-driven decision-making processes, smart cities can better address the diverse needs and aspirations of their residents. Moreover, this book has showcased successful examples of citizen engagement initiatives, where citizens are empowered to actively participate in shaping the development and implementation of smart city projects.

The previous considerations have underscored the significance of long-term planning and sustainable urban development in the smart city landscape. In the previous chapter, I emphasized the need to consider social and environmental sustainability principles in the design and implementation of smart city projects. By adopting a holistic approach that prioritizes inclusivity, environmental preservation, and resource efficiency, smart cities can foster livable and resilient urban environments for current and future generations.

The key takeaways and insights gained from the exploration of smart cities in the digital age highlight the importance of collaboration, inclusivity, and long-term planning for the successful development and management of smart cities. The challenges and opportunities discussed throughout this book have underscored the need for stakeholders to work together, leverage technological innovations responsibly, and prioritize the well-being and aspirations of citizens.

Thus, the Global Smart City: Challenges and Opportunities in the Digital Age provides a comprehensive guide for readers to understand the complexities and dynamics of smart cities. By delving into the various dimensions of smart city initiatives, this book equips policymakers, urban planners, industry professionals, and citizens with the knowledge and insights necessary to navigate the rapidly evolving landscape of smart cities. I hope that this book inspires collaboration, sparks innovative ideas, and contributes to the creation of sustainable, inclusive, and prosperous smart cities in the digital age. The upcoming section will provide valuable suggestions on successfully navigating the smart city landscape, offering practical insights for policymakers, managers, citizens, and users. These suggestions will help stakeholders effectively navigate the complex challenges and seize the numerous opportunities presented by smart cities.

8.2 Guideline for Navigating the Smart City Edge

This book has delved into the various internal facets of the smart city landscape and the impact it has on individuals, businesses, tourists, and urban development. In today's rapidly evolving digital age, effectively navigating the smart city landscape is of utmost importance. It requires a comprehensive understanding of the diverse roles played by citizens, users, managers, and policymakers in shaping the future of urban environments. By adopting a citizen-centric approach and fostering collaboration among various stakeholders, smart cities can fully harness the potential of digital technologies to improve quality of life, promote sustainability, and drive innovation.

As smart cities continue to evolve, it becomes crucial for all stakeholders to proactively comprehend the ongoing changes and adapt accordingly. This section aims to provide guidelines for success in navigating the smart city landscape, focusing on two key areas: empowering citizens and users, and strategies for managers and policymakers. By offering insights from both current and future perspectives, this book aims to equip readers with valuable knowledge on how to thrive and flourish within the dynamic smart city ecosystem.

8.2.1 Citizens and Users

Cities around the world are undergoing a remarkable shift as they embrace digital advancement, sustainability, livability, and efficiency as integral components of the smart city movement. This transformative process highlights the importance for citizens to recognize their pivotal role and potential in shaping the future of their urban environments. It is imperative that citizens understand the significance of their engagement and contribution to the development of smart cities. This section aims to provide citizens with practical guidelines and insightful suggestions to effectively navigate the dynamic ecosystem of smart cities, considering both current realities and future prospects. As previously discussed, smart city initiatives are based on two approaches: top-down and bottom-up. These approaches result in different interactions between citizens and institutions. The top-down approach involves a centralized decision-making process driven by city authorities and stakeholders. On the other hand, the bottom-up approach emphasizes the active participation of citizens in shaping their cities' development. These approaches are not mutually exclusive; rather, they complement each other, forming a symbiotic relationship that realizes the full potential of smart cities.

Focusing on the bottom-up approach, it has been recognized that citizens play a central role in shaping and contributing to the development of their cities. This approach, rooted in Europe and the United States, highlights citizens' active involvement in driving the adoption of smart city technologies. Technological advancements enable citizens to actively participate in data collection, analysis, and decision-making processes. This fosters a sense of ownership, responsibility, and innovation among citizens, transforming them into key stakeholders in smart city development. The successful implementation of citizen-centric approaches in cities like Amsterdam serves as a remarkable example as we have seen in Chapter 4.

Today, citizens are the lifeblood and the engine of the city, and their ability to adapt to a dynamic and unfamiliar environment is crucial. The COVID-19 pandemic has demonstrated that citizens, regardless of their technological readiness, can perceive and participate in the process of digitalization of smart cities. This reaction is an important indicator of how humans are predisposed to adapt and modify their habits and societal views when accompanied by the evolving world around them.

This concept leads to the notion of citizen resilience, defined as the ability to adapt to change and overcome adversity. Resilience is crucial in the development of smart cities, as citizens must be willing to open themselves up to change and modify their roles in the urban environment. However, this change cannot happen alone and requires the involvement of the state and institutions, as discussed in the next subsection.

Therefore, in order to enhance their central position in today's cities, citizens need to carefully navigate the diverse dynamics, benefits, and challenges associated with embracing the transformation of urban landscapes. They must effectively navigate the intricate intersections of technological, social, and cultural evolutions, while understanding the implications and opportunities that arise from this paradigm shift.

To address these challenges and leverage the benefits of transforming urban landscapes, the following guidelines and suggestions are proposed to assist citizens in effectively navigating the complex smart city ecosystem:

- *Bridging the Technological Divide*
  Bridging the Technological Divide is a crucial step in empowering citizens to actively participate in the smart city transformation. Both the city and its residents have important roles to play in this process.

  From the city's perspective, it is essential to provide accessible resources and initiatives that facilitate technological education and training for citizens. This can be achieved through the establishment of training programs, workshops, and digital literacy campaigns. By investing in these initiatives, cities can ensure that citizens have the necessary skills and knowledge to effectively engage with digital tools and technologies. In this line, present and future citizens have a responsibility and a need to bridge the technological divide and enhance their digital skills and knowledge. Clearly, this transformation currently encompasses different age groups, each with varying degrees of responsiveness to technology. However, the pandemic has demonstrated how these practices have the potential to overcome resistance and bring users closer to embracing this transformation. Undoubtedly, the gap between citizens and users must be reduced. In a futuristic but not distant vision, which is already partially underway, citizens should adopt digital tools and technologies in their daily lives and explore their potential to improve various aspects of urban living.

  An example of a city successfully bridging the technological divide is Barcelona, Spain. The city has implemented a comprehensive digital inclusion strategy that aims to provide equal access to technology for all citizens.

Barcelona offers free Wi-Fi hotspots in public spaces, provides digital training programs for residents, and promotes the use of digital platforms to enhance citizen participation in decision-making processes (Bakici, Almirall, & Wareham, 2013; Bourliataux-Lajoinie, Dosquet, & del Olmo Arriaga, 2019). By prioritizing technological accessibility, Barcelona has empowered its citizens to actively participate in shaping the city's smart future.

Thus, bridging the technological divide requires collaborative efforts from both the city and its citizens. By providing accessible resources and training opportunities, cities can empower citizens to embrace digital tools and technologies. Simultaneously, citizens must actively seek out these opportunities, enhance their digital skills, and adopt digital solutions in their daily lives.

- *Embracing a New Vision of the City*

  To fully embrace a new vision of the city, citizens must be receptive to a paradigm shift toward a more efficient and digitally driven urban environment. This transformation may significantly deviate from the familiar traditional urban landscape, necessitating a willingness to adapt and embrace change. By embracing innovative concepts and technologies, citizens hold the power to actively contribute to the creation of a sustainable, inclusive, and livable urban environment (Christofi, Iaia, Marchesani, & Masciarelli, 2021; Dirks & Keeling, 2009; Linde, Sjödin, Parida, & Wincent, 2021; Tiwana, 2014; Vanolo, 2014).

  In order to achieve this objective, citizens should acknowledge that building a smart city necessitates collective effort and engagement. By forming community networks, neighborhood associations, or participating in local initiatives, citizens can foster collaboration, social cohesion, and overall well-being among residents. Through collective action, citizens can leverage their combined knowledge, skills, and resources to shape the city in accordance with their shared values and aspirations.

  Additionally, promoting and embracing inclusivity and accessibility is of paramount importance to ensure that the benefits of the smart city transformation are accessible to all citizens, regardless of their socioeconomic status, age, or abilities. Along these lines, citizens should adopt environmentally friendly behaviors, such as promoting energy efficiency, reducing waste, and supporting sustainable transportation options. By embracing sustainability, citizens can contribute to a greener, healthier, and more resilient urban environment.

  Consequently, actively contributing to a new vision of the city requires citizens to be dynamic, resilient, and adaptable. By embracing the new vision and actively participating in the city's transformation, citizens have the power to shape their urban environment and create a brighter future for themselves and future generations.

  Citizens need to be open to a new vision of a more efficient and digitally driven city, which may differ from the traditional urban landscape they are accustomed to. By embracing innovative concepts and technologies, citizens can contribute to creating a sustainable and livable urban environment. This entails embracing change and adapting to new ways of engaging with the city and its services.

- *Actively Shaping City Trajectories*
  Citizens hold a significant role in shaping the future trajectory of their cities. To effectively contribute, it is crucial for citizens to actively engage in various avenues of local and digital participation. By voicing their opinions, concerns, and ideas, citizens can exert influence over the decision-making processes that shape the development plans and policies of the smart city.
  In this line, smart city projects and digital services in city allow citizens and users to explore opportunities for involvement in urban planning processes. This can include participating in public consultations organized by city authorities or contributing to the development of local master plans. By actively engaging in these processes, citizens can share their insights and ideas, providing valuable input that helps shape the future direction of the city. Moreover, embedded in the smart city nature, a key point for citizen participation is through the use of digital platforms and technology. Many cities now provide online platforms where citizens can provide feedback, propose ideas, and engage in discussions on various urban issues. Citizens can leverage these platforms to contribute their unique perspectives and innovative solutions, ensuring that diverse voices are heard in the decision-making process (Abid, Marchesani, Ceci, Masciarelli, & Ahmad, 2022; Benevolo, Dameri, & Auria, 2016; Gil-Garcia, Zhang, & Puron-Cid, 2016; Pan, Lin, Chuang, & Kao, 2011).
  An exemplary case of citizen-driven initiatives can be seen in the city of Barcelona previously discussed, where citizens actively participated in the development of the "Superblocks" project. Through public consultations and collaboration with local residents, the city transformed certain areas into car-free zones, prioritizing pedestrian spaces and promoting community engagement. This project demonstrates how citizen involvement can shape the urban environment to align with their vision of a sustainable, inclusive, and people-centric city.
  Therefore, it is assumed that citizens possess the agency and capability to actively influence the trajectory of their cities. By embracing and actively participating in the various opportunities for engagement, citizens are empowered to cocreate the urban environment they desire. This entails directly participating in the decision-making processes and actively contributing to the creation and development of the cities of the future.

- *Leveraging Smart City Services for Personal Well-being*
  Citizens have the opportunity to enhance their well-being and quality of life by making the most of the wide range of services and developments offered by smart cities. These advancements, which encompass areas such as health care, transportation, and energy, can greatly contribute to improving daily experiences and overall satisfaction within the urban environment.
  For example, in a smart city, health-care services are integrated with technology to provide efficient and personalized care. Citizens can take advantage

of digital health platforms that enable remote consultations, access to medical records, and real-time health monitoring. By embracing these services, individuals can proactively manage their health, receive timely medical attention, and enjoy a higher quality of life. For instance, wearable devices can track vital signs and provide personalized health recommendations, promoting a healthier lifestyle.

Another perspective considers the efficient transportation networks are another key element of smart cities that citizens can leverage for their well-being. Smart transportation solutions offer convenience, reduced congestion, and improved safety. Citizens can explore various options such as ride-sharing, electric vehicles, and intelligent traffic management systems to optimize their commuting experience. By choosing sustainable modes of transportation, citizens can contribute to reducing air pollution, easing traffic congestion, and creating a more livable urban environment. For example, utilizing smart public transportation apps can provide real-time information on routes, schedules, and traffic conditions, allowing individuals to plan their journeys efficiently (Abid et al., 2022; Chu, Cheng, & Yu, 2021; Gil-Garcia et al., 2016; Mangiaracina, Perego, Salvadori, & Tumino, 2017).

Furthermore, smart cities emphasize sustainable energy solutions that citizens can tap into for their personal well-being. Renewable energy sources, smart grid systems, and energy-efficient buildings are integral parts of a smart city's infrastructure. Citizens can adopt energy-saving practices in their homes, such as utilizing smart thermostats, solar panels, and energy-efficient appliances (Dobbelsteen et al., 2020; Massana, Pous, Burgas, Melendez, & Colomer, 2017). These actions not only contribute to environmental sustainability but also help individuals save on energy costs and create a comfortable living environment. For instance, smart energy management systems can monitor energy consumption patterns, identify areas for improvement, and provide personalized recommendations for reducing energy usage.

In order to fully leverage smart city services for personal well-being, citizens should familiarize themselves with the available technologies and actively engage with them. It is important to stay informed about new developments, services, and initiatives through official city communication channels, community events, and online platforms. Citizens can also participate in workshops, training programs, and educational campaigns organized by the city or community organizations to enhance their digital literacy and technological skills. By doing so, individuals can effectively navigate the smart city ecosystem, optimize their usage of smart services, and maximize the benefits for their well-being.

- *Seizing Opportunities for Active Navigation*

  Citizens and users play a central and indispensable role in the ongoing transformation taking place within cities. It is crucial to acknowledge their potential to actively navigate and shape the urban landscape. As we embrace the advent of the digital era and avail ourselves of the myriad offerings presented by smart cities, individuals have the opportunity to revolutionize their interactions with the urban environment, transcending conventional boundaries and elevating their overall urban experience. This necessitates harnessing the power of digital platforms, mobile applications, and intelligent infrastructure to unlock real-time information, seamless transportation networks, and immersive engagement with city services (Gagliardi et al., 2017; Marchesani, 2022; Marchesani, Masciarelli, & Bikfalvi, 2023; Nam & Pardo, 2014).

At the core of this transformative journey lies the effective utilization of digital platforms and mobile applications. Smart cities have become veritable repositories of digital resources, waiting to be explored and harnessed by citizens. By immersing themselves in the expansive realm of mobile applications, individuals can unlock a wealth of services and information at their fingertips. These applications encompass a wide array of functionalities, ranging from transit planning and navigation aids that unravel the intricate web of urban mobility, to cultural event guides that uncover the city's vibrant arts scene, and community engagement platforms that foster meaningful connections and collective participation. Through these digital tools, citizens can emerge as active participants, shaping the city's narrative, and leaving an indelible mark on its fabric.

Furthermore, by comprehending and enhancing the relationship between citizens and intelligent infrastructures, citizens and users have directly influenced and been affected by the trajectories of cities.

Sensors, IoT devices, and interconnected networks form the foundation of a data-driven urban landscape. Citizens can seize this transformative potential by tapping into the rich tapestry of real-time data and analysis (Gaur, Scotney, Parr, & McClean, 2015; B. N. Silva, , Khan, & Han, 2018; A. Wang et al., 2020). Imagine deciphering traffic patterns to chart the most efficient routes, monitoring air quality to make informed decisions about outdoor activities, and tracking energy consumption to contribute to a more sustainable future. Equipped with these invaluable insights, individuals can curate their urban journeys, optimizing their routines, and ultimately reshaping the urban landscape to foster a more efficient, environmentally conscious, and harmonious coexistence.

In conclusion, citizens are implored to embrace the tantalizing opportunities and limitless possibilities that cities offer, inviting them to embark on an extraordinary voyage of active navigation and personal growth within the urban landscape. By unlocking the potential of digital platforms, mobile applications, and intelligent infrastructure, individuals transcend the boundaries of conventional urban experiences. Through these channels, citizens become architects of their urban destinies, forging paths that reflect their aspirations and values, and contributing to a legacy that resonates throughout their communities. Let us embark on this transformative journey together, where the city becomes our canvas, and our collective actions paint a portrait of progress, ingenuity, and fulfillment.

8.2.2 Managers and Policymakers

As cities embrace the concept of smart cities and embark on the journey of digital transformation, the role of public managers and policymakers becomes paramount in ensuring successful implementation. The complexities of smart city initiatives require effective leadership and strategic decision-making to navigate the multifaceted challenges and opportunities that arise. Public managers and policymakers must take a proactive approach in orchestrating these dynamics and understanding the context in which smart city implementation takes place.

First and foremost, managers and policymakers should develop a comprehensive understanding of their city's unique context and challenges. Each city has its own social, economic, and cultural characteristics that shape the implementation of smart city initiatives. By conducting thorough research, engaging with stakeholders, and conducting needs assessments, managers and policymakers can identify the specific needs and aspirations of their citizens. This contextual understanding serves as a foundation for developing tailored strategies and solutions that address the specific challenges faced by the city. Effective governance is a critical component of successful smart city implementation. Managers and policymakers should establish clear governance structures that define roles, responsibilities, and decision-making processes (Abid et al., 2022; Bakici et al., 2013; Karimikia et al., 2022; K. A. Paskaleva, 2009; Pittaway & Montazemi, 2020; Soares, Domareski Ruiz, & Ivars Baidal, 2021). This includes identifying key stakeholders and fostering collaboration among them. Engaging with citizens, community organizations, businesses, and academia allows for a diversity of perspectives and ensures that the decision-making process is inclusive and participatory. By involving all relevant stakeholders, managers and policymakers can ensure that the smart city initiatives align with the needs and expectations of the community.

Furthermore, managers and policymakers should prioritize the development of policies and regulations that promote transparency, accountability, and ethical use of data. Smart cities rely on data collection and analysis to enhance services and improve quality of life. However, it is crucial to establish robust data governance frameworks that protect citizen privacy, ensure data security, and address potential biases or discriminatory practices. By establishing guidelines for data usage, managers and policymakers can build trust with citizens and foster a positive perception of smart city initiatives. To effectively engage with and address the needs of citizens, managers and policymakers should adopt a citizen-centric approach. This involves actively involving citizens in decision-making processes and seeking their input throughout the implementation journey. Public managers and policymakers can organize workshops, town hall meetings, and online platforms to gather feedback and insights from citizens. This inclusive approach not only fosters a sense of ownership and empowerment among citizens but also ensures that smart city initiatives are designed to meet their specific needs.

Moreover, effective communication is vital in ensuring the success of smart city implementation. Managers and policymakers should employ clear and transparent communication strategies to disseminate information about the objectives, progress, and benefits of smart city initiatives. This includes regular updates, public awareness campaigns, and educational programs to ensure that citizens are well-informed and understand the value of these initiatives. By promoting open communication channels, managers and policymakers can build trust, manage expectations, and address any concerns or misconceptions that may arise. Additionally, cities can collaborate with educational institutions, community organizations, and private sector entities to create partnerships that promote technology accessibility. For instance, they can establish public–private partnerships to provide free or low-cost access to digital devices, internet connectivity, and online learning platforms. By making these resources readily available, cities can bridge the technological gap and empower citizens to participate actively in the smart city ecosystem.

One of the key recommendations of this book is for managers and policymakers to adopt a holistic and integrated approach when designing and implementing smart city initiatives. This involves considering the interdependencies between various urban systems, as well as the needs and preferences of different stakeholders. By doing so, they can develop comprehensive strategies that address multiple challenges simultaneously and create synergies between different initiatives, ultimately leading to more efficient and sustainable urban development. To successfully navigate smart city implementation, public managers and policymakers should consider several key aspects.

First, fostering collaboration and cross-sector partnerships is essential. This involves encouraging cooperation among different departments, agencies, and sectors involved in smart city initiatives, leveraging expertise from urban planning, transportation, energy, and public safety. Establishing platforms for knowledge sharing promotes a multidisciplinary approach.

Second, embracing data integration and interoperability is crucial. Developing data governance frameworks that facilitate seamless data exchange between urban systems and services, utilizing open standards and APIs, and establishing data sharing agreements prioritizing privacy and security enables data-driven decision-making.

Third, designing citizen-centric solutions is paramount. Placing citizens at the center of decision-making, conducting user research, and engaging with citizens to understand their needs, preferences, and concerns allows for the creation of solutions that address their priorities, enhance their quality of life, and promote inclusivity and social equity.

Considering long-term sustainability is also essential. Incorporating sustainability principles into smart city initiatives, such as considering environmental, social, and economic factors, integrating renewable energy sources, energy-efficient technologies, and sustainable transportation solutions, ensures the long-term viability and resilience of urban systems.

Fostering innovation and entrepreneurship is another vital aspect. Creating an environment that encourages innovation, supporting the development of innovation hubs, incubators, and accelerators, and facilitating public–private partnerships leverages private sector expertise, investment, and innovation in smart city projects.

Lastly, engaging in continuous evaluation and adaptation is crucial. Establishing mechanisms for ongoing evaluation and monitoring of smart city initiatives, collecting and analyzing data to assess their impact, effectiveness, and efficiency, and using evaluation findings to inform future decision-making, adapt strategies, and improve service delivery are essential for successful implementation.

By considering these recommendations, public managers and policymakers can navigate the complexities of smart city implementation and contribute to the development of sustainable, citizen-centric urban environments.

To address these challenges and leverage the benefits of transforming urban landscapes, the following guidelines and suggestions are proposed to assist public managers and policymakers in effectively navigating the complex smart city ecosystem focusing on a clear relationship with urban environment, users, and citizens:

- *Fostering Cross-Sector Collaboration*

  To navigate the smart cities of the future successfully, public managers and policymakers must prioritize the fostering of cross-sector collaboration. The integration of different sectors and stakeholders is vital for the success of smart city projects. This book strongly encourages managers and policymakers to take proactive steps in promoting collaboration and creating a conducive environment for cross-sector partnerships.

  One key recommendation is the establishment of platforms for dialogue and knowledge sharing. These platforms serve as spaces for stakeholders from public, private, and nonprofit organizations (NPOs) to come together, exchange ideas, and share best practices. By facilitating regular communication and collaboration, managers and policymakers can foster a collaborative ecosystem that harnesses the collective resources, expertise, and innovative potential of various stakeholders.

  Furthermore, managers and policymakers should actively seek to establish partnerships between different sectors. Collaborating with diverse stakeholders, including government agencies, businesses, research institutions, and community organizations, can unlock synergies and create a multiplier effect in smart city initiatives.

  These partnerships enable the pooling of resources, the sharing of knowledge, and the collective pursuit of common goals.

  In navigating the complexities of smart cities, managers and policymakers must recognize that no single entity possesses all the necessary knowledge and resources. By embracing cross-sector collaboration, they tap into a wealth of expertise and perspectives, enabling them to develop comprehensive strategies and implement holistic solutions. However, it is important for managers and policymakers to be aware of the challenges that may arise in fostering cross-sector collaboration. Different sectors often have distinct objectives, priorities, and decision-making processes. It is crucial to establish effective communication channels and mechanisms for resolving conflicts or differences in interests. Additionally, the creation of a collaborative environment requires nurturing trust, promoting open dialogue, and creating a shared vision for the future of the smart city (Bernardi & Diamantini, 2018; Cohen, Almirall, & Chesbrough, 2016; Meijer & Bolívar, 2016; Nam & Pardo, 2014; Salom-Carrasco, Pitarch-Garrido, & Sales-Ten, 2017; Trencher, 2019).

Thus, it assumes that fostering cross-sector collaboration is an essential aspect of navigating the smart cities of the future. By creating platforms for dialogue, sharing best practices, and establishing partnerships, managers and policymakers can leverage the collective knowledge, resources, and innovative capabilities of diverse stakeholders. This collaborative approach will contribute to the development of more effective, sustainable, and citizen-centric smart city initiatives.

- *Ensuring Data Security and Privacy*
  In the era of smart cities, where data collection and analysis are integral to improving urban environments, public managers and policymakers must prioritize the protection of data security and privacy. This book emphasizes the critical need for managers and policymakers to develop robust data governance frameworks that safeguard citizens' privacy and ensure responsible data management practices.

  To navigate the smart cities of the future successfully, managers and policymakers should adopt stringent measures to protect data security. This includes implementing strong encryption methods, employing secure data storage protocols, and establishing robust access controls. By employing these safeguards, they can mitigate the risk of unauthorized access, data breaches, and cyber threats, safeguarding the privacy and confidentiality of citizens' personal information. Transparency in data management is another essential aspect of ensuring data security and privacy. Managers and policymakers should establish clear data usage policies that define how data is collected, stored, and utilized within the smart city ecosystem. These policies should also outline citizens' rights regarding their data, including consent, access, and the ability to control the use of their information. By promoting transparency, managers and policymakers can build trust and confidence among citizens, fostering a positive perception of smart city initiatives. Moreover, it is crucial for managers and policymakers to ensure the ethical use of data in smart city applications. This involves establishing guidelines and best practices for data collection, analysis, and sharing that align with privacy regulations and respect individual rights. It is also essential to address potential biases and discriminatory practices that may arise from data-driven decision-making. By adopting a proactive approach to ethical data management, managers and policymakers can mitigate risks and ensure that smart city initiatives benefit all citizens equitably.

  Finally, to effectively navigate the complexities of data security and privacy, collaboration and partnerships between public and private entities are vital. Managers and policymakers should engage with experts in cybersecurity and data privacy to stay updated on evolving threats and best practices. Collaboration with industry leaders, academic institutions, and research organizations can also contribute to the development of innovative solutions and frameworks that enhance data security and privacy in smart city initiatives.

In conclusion, ensuring data security and privacy is a fundamental aspect of navigating the smart cities of the future. By developing robust data governance frameworks, promoting transparency, and fostering ethical data practices, managers and policymakers can build trust, protect citizens' privacy, and enable the responsible and beneficial use of data in smart city applications. By prioritizing data security and privacy, they can create a foundation of trust and confidence that is essential for the success and acceptance of smart city initiatives.

- *Encouraging Innovation and Experimentation*
In the rapidly evolving landscape of smart cities, innovation and experimentation play a crucial role in driving progress and shaping the future of urban development. This book emphasizes the significance of fostering an environment that encourages innovation and experimentation, and it offers recommendations to managers and policymakers on how to navigate this dynamic landscape effectively.

To navigate the smart cities of the future successfully, managers and policymakers should embrace a culture of innovation. They can achieve this by supporting research and development initiatives that explore emerging technologies and innovative solutions for urban challenges (B. N. Silva et al., 2018; Supangkat, Arman, Nugraha, & Fatimah, 2018; Wicaksana, 2020). By allocating resources and funding to support these efforts, they can create an ecosystem that encourages creative problem-solving and the development of cutting-edge smart city solutions. Furthermore, managers and policymakers should facilitate experimentation by providing opportunities for pilot projects and testbeds. These initiatives allow for the exploration and validation of new ideas and technologies on a smaller scale before their full-scale implementation. By creating spaces and regulatory frameworks that support experimentation, managers and policymakers can enable the identification of successful approaches and the rapid iteration and improvement of smart city solutions.

To encourage innovation and experimentation, collaboration is essential. Managers and policymakers should establish partnerships with academia, research institutions, and private sector entities to leverage their expertise and resources. By fostering collaboration, they can tap into a diverse range of perspectives, knowledge, and skills, promoting the cocreation of innovative solutions that address the specific needs and aspirations of their cities. Additionally, managers and policymakers should facilitate knowledge sharing and learning within their smart city ecosystem (Adler & Florida, 2021; Freeman, 2010; Kummitha & Crutzen, 2019; Marchesani et al., 2023). They can organize conferences, workshops, and networking events that bring together key stakeholders, fostering the exchange of ideas, experiences, and best practices. By promoting a culture of learning and collaboration, they can create an environment where innovation thrives and where stakeholders can collectively learn from both successes and failures.

It is important for managers and policymakers to create an agile regulatory environment that embraces innovation and adapts to technological advancements. They should establish frameworks that are flexible and adaptive, allowing for the integration of new technologies and business models into the fabric of the city. By creating a regulatory environment that promotes innovation rather than stifles it, managers and policymakers can attract investment, talent, and entrepreneurs to their cities, fostering an ecosystem of innovation and economic growth.

Thus, it assumes that encouraging innovation and experimentation is paramount for navigating the smart cities of the future. By supporting research and development, facilitating pilot projects, fostering collaboration, and creating an agile regulatory environment, managers and policymakers can cultivate a culture of innovation that drives the continuous improvement of smart city solutions. By embracing innovation, they can ensure that their cities remain at the forefront of technological advancements and effectively address the evolving needs and aspirations of their citizens.

- *Prioritizing Sustainability and Resilience*
  In the face of mounting environmental challenges and the need for long-term urban development, prioritizing sustainability and resilience is crucial for the success of smart city initiatives. This book highlights the importance of integrating environmental, social, and economic considerations into the fabric of smart cities and offers recommendations to managers and policymakers on how to effectively navigate this critical aspect.

  To navigate the smart cities of the future, managers and policymakers should prioritize sustainability by adopting a holistic approach that encompasses environmental, social, and economic dimensions. They can achieve this by integrating sustainable practices into the planning, design, and implementation of smart city projects. This includes promoting resource efficiency, reducing greenhouse gas emissions, and enhancing the use of renewable energy sources. By prioritizing sustainable development, they can minimize the ecological footprint of their cities and contribute to global climate action goals (Abid et al., 2022; Contreras & Platania, 2019; Khatoun & Zeadally, 2016; Noori, de Jong, Janssen, Schraven, & Hoppe, 2021; Zhu, Li, & Feng, 2019).

  Resilience is equally important in ensuring the long-term viability of smart cities. Managers and policymakers should prioritize the resilience of urban systems by enhancing their capacity to withstand and recover from shocks and stresses. This involves designing infrastructure and services that are adaptable, flexible, and robust, capable of withstanding natural disasters, cyberattacks, and other potential disruptions. By prioritizing resilience, they can safeguard the well-being of their citizens and protect critical urban assets. To prioritize sustainability and resilience, managers and policymakers should promote collaboration among stakeholders, including government agencies, businesses, community organizations, and research institutions. By fostering collaboration, they can leverage collective expertise and resources to develop innovative solutions that address sustainability challenges and enhance resilience. Collaboration also facilitates the sharing of best practices and lessons learned, enabling cities to learn from each other and accelerate progress toward sustainable and resilient futures.

Additionally, managers and policymakers should integrate sustainability and resilience criteria into the decision-making processes for smart city projects. This involves conducting comprehensive assessments that consider environmental impacts, social equity, and economic viability. By incorporating these criteria, they can ensure that projects align with broader sustainability goals and deliver long-term benefits to their cities and citizens. Moreover, education and awareness play a vital role in prioritizing sustainability and resilience. Managers and policymakers should promote public engagement and education initiatives that raise awareness about the importance of sustainability and resilience in smart cities. By empowering citizens with knowledge and fostering a sense of ownership, they can encourage individual actions and community participation in achieving sustainable and resilient urban futures.

Thus, it assumes that prioritizing sustainability and resilience is essential for navigating the smart cities of the future. By integrating sustainable practices, enhancing resilience, fostering collaboration, incorporating sustainability and resilience criteria in decision-making processes, and promoting public awareness, managers and policymakers can ensure the long-term viability and well-being of their cities. By prioritizing sustainability and resilience, they can create smart cities that not only meet the needs of the present but also contribute to a sustainable and resilient future for generations to come.

- *Evaluating and Monitoring the Impact of Smart City Initiatives*

  The evaluation and monitoring of smart city initiatives are essential for managers and policymakers to gauge their effectiveness, identify areas for improvement, and make informed decisions for future development. This book emphasizes the importance of adopting a systematic approach to assess the impact of smart city projects and provides recommendations to guide managers and policymakers in effectively navigating this aspect.

  To navigate the smart cities of the future, managers and policymakers should prioritize the evaluation and monitoring of smart city initiatives by establishing robust performance metrics and indicators. These metrics should align with the goals and objectives of the projects and capture both quantitative and qualitative aspects of their impact. By defining clear and measurable indicators, they can assess the progress, outcomes, and benefits of smart city initiatives with accuracy and consistency. Regular evaluation allows managers and policymakers to track the performance of smart city projects over time and compare them against predetermined benchmarks or targets. By conducting periodic assessments, they can identify strengths, weaknesses, and areas that require improvement (Chauhan, Jakhar, & Chauhan, 2021; G. Li, Wang, Luo, & Li, 2018; Orlowski & Romanowska, 2019; K. Paskaleva & Cooper, 2018; Ullah, Qayyum, Thaheem, Al-Turjman, & Sepasgozar, 2021; Yan, Liu, &

Tseng, 2020). This iterative process enables them to make data-driven decisions, optimize resource allocation, and enhance the efficiency and effectiveness of smart city initiatives. Moreover, monitoring plays a complementary role in the evaluation process by providing real-time data and insights on the ongoing performance of smart city projects. By implementing monitoring mechanisms, such as data collection systems, sensors, and feedback loops, managers and policymakers can gather timely information on key performance indicators. This continuous monitoring enables them to identify emerging issues, detect deviations from expected outcomes, and take proactive measures to address any challenges or obstacles.

To effectively evaluate and monitor smart city initiatives, managers and policymakers should consider the multidimensional nature of their impact. This includes assessing environmental, social, economic, and governance dimensions to capture a comprehensive understanding of the benefits and implications of these projects. By adopting a holistic approach to evaluation, they can ensure a balanced assessment that reflects the diverse perspectives and interests of stakeholders. Furthermore, managers and policymakers should leverage data analytics and emerging technologies to enhance the evaluation and monitoring process. By harnessing the power of data, they can uncover patterns, trends, and insights that inform decision-making and drive continuous improvement. This may involve the use of advanced analytics, machine learning (ML), and AI techniques to analyze large datasets and extract meaningful information.

Last but not least, it is essential to communicate the findings of the evaluation and monitoring efforts to stakeholders, including citizens, community organizations, and businesses. By sharing the results openly and transparently, managers and policymakers can foster trust, solicit feedback, and demonstrate their commitment to evidence-based decision-making. Effective communication of evaluation findings also facilitates learning, knowledge sharing, and collaboration among stakeholders.

Thus, it assumes that evaluating and monitoring the impact of smart city initiatives is crucial for managers and policymakers to make informed decisions, optimize resources, and ensure the continuous improvement of their cities. By adopting a systematic approach, establishing performance metrics, leveraging data analytics, engaging external evaluators, and communicating findings, they can navigate the complexities of smart city implementation and pave the way for sustainable and resilient urban development.

- *Lending to an entrepreneurial vision of smart city:*
  Finally, as a perspective guideline for public managers and policymakers, the discussion emerging from this book suggests that cities should embrace an entrepreneurial vision that encourages innovation, fosters collaboration, and stimulates economic growth. Specifically, this book highlights the importance of creating an environment that supports entrepreneurship and provides guidelines for managers and policymakers to effectively navigate this aspect of smart city development.

First and foremost, fostering an environment that encourages entrepreneurship and innovation is crucial. Managers and policymakers should establish policies and initiatives that promote a culture of entrepreneurship within the smart city ecosystem. This involves creating a supportive infrastructure that nurtures start-ups and small businesses, such as innovation hubs, incubators, and accelerators. By providing access to resources, mentorship, and networking opportunities, they can facilitate the growth and success of entrepreneurial ventures.

In addition to supporting the development of start-ups, managers and policymakers should encourage public–private partnerships and collaboration. Collaborative initiatives between government entities, private sector organizations, and academia can leverage the expertise, resources, and innovation potential of different stakeholders. By fostering collaboration, they can create synergies, share knowledge, and drive the cocreation of smart city solutions (Allam & Jones, 2020; Kummitha, 2018, 2020; Kummitha & Crutzen, 2017; Marchesani et al., 2023; Marchesani, Iaia, Masciarelli, & Christofi, 2022; Sharifi, Khavarian-Garmsir, & Kummitha, 2021). This collaborative approach also enhances the scalability and sustainability of entrepreneurial initiatives, leading to a more vibrant and dynamic smart city ecosystem. Creating incentives and funding opportunities for entrepreneurial initiatives is another crucial step. Managers and policymakers should establish programs and mechanisms that provide financial support, grants, and incentives for start-ups and small businesses in the smart city sector. These incentives can include tax breaks, access to funding, and streamlined regulatory processes. By lowering barriers to entry and providing financial incentives, they can attract entrepreneurial talent and stimulate the development of innovative solutions that address the unique challenges of their cities.

Moreover, managers and policymakers should emphasize the economic opportunities and job creation potential of smart city technologies and solutions. By showcasing the economic benefits of adopting smart city initiatives, such as increased productivity, job creation, and enhanced competitiveness, they can garner support from various stakeholders, including businesses, investors, and the public. Highlighting the economic value proposition of smart city development can attract private sector investment, stimulate economic growth, and create employment opportunities, ultimately contributing to the overall prosperity of the city. To effectively navigate the entrepreneurial dimension of smart city development, managers and policymakers should also stay informed about emerging technologies, trends, and market dynamics. By continuously monitoring and understanding the evolving landscape of the smart city ecosystem, they can adapt policies, regulations, and support mechanisms to foster an environment conducive to entrepreneurship. This includes staying updated on technological advancements, engaging with industry experts and thought leaders, and actively participating in global networks and knowledge-sharing platforms.

Lastly, managers and policymakers should communicate the importance of entrepreneurship and the value of smart city initiatives to the public and other stakeholders. By raising awareness and promoting understanding of the benefits and opportunities associated with entrepreneurship in the smart city context, they can generate support, engagement, and buy-in from the community. Effective communication and public engagement strategies can help build trust, manage expectations, and create a shared vision for the entrepreneurial future of the smart city.

Thus, it assumes that embracing an entrepreneurial vision is essential for navigating the complexities of smart city development. By fostering an environment that encourages entrepreneurship, supporting start-ups and small businesses, promoting collaboration, providing incentives, emphasizing economic opportunities, staying informed about emerging trends, and communicating effectively, managers and policymakers can create a thriving entrepreneurial ecosystem that drives innovation, economic growth, and the sustainable development of smart cities.

8.3 Smart City Limits and Criticism

Smart city initiatives have the potential to deliver significant benefits, but they also face several limitations and criticisms. To ensure the successful implementation of smart city projects, it is important to address these concerns from the perspectives of both citizens and managers/policymakers.

One of the key challenges is addressing the digital divide, which refers to the unequal access to digital technologies and the internet. Marginalized communities, including low-income neighborhoods, people with disabilities, and the elderly, may face barriers in accessing and utilizing digital tools and services. This digital divide can further deepen existing disparities in access to essential services, education, employment opportunities, and civic participation. To promote social cohesion and reduce these disparities, inclusive smart city initiatives must prioritize the needs of vulnerable populations. This requires adopting strategies that ensure equitable access to digital infrastructure, affordable connectivity, and user-friendly interfaces.

Inclusive smart city planning should also consider the specific needs and challenges faced by different segments of society. For example, ensuring accessible transportation options for people with disabilities, implementing inclusive design principles in the built environment, and offering digital literacy training programs for elderly residents can foster inclusivity and ensure that no one is left behind in the digital transformation of cities. By actively addressing these issues, smart cities can bridge the gap between different segments of society and promote a more just and equitable urban future. Inclusive smart city initiatives not only provide equal opportunities and access to services for all residents but also promote social cohesion, community engagement, and economic empowerment. Moreover, it is important to incorporate indicators of social equity and inclusivity into the evaluation and monitoring of smart city initiatives can help identify any unintended consequences or disparities. By regularly assessing the impact of these initiatives on different segments of society, managers and policymakers can make data-driven decisions and adjust strategies as needed to address any emerging inequities. By adopting strategies that address the specific needs of vulnerable populations, engaging with community organizations, and continuously monitoring the impact of these initiatives, managers and policymakers can ensure that smart cities become engines of equitable and sustainable urban development.

This discussion opens up the ethical debate surrounding smart cities and contributes to the ongoing discourse among academics, urban planners, and economists (Berrone, Ricart, & Carrasco, 2016; Datta, 2015; Datta & Odendaal, 2019; Grossi & Pianezzi, 2017; Hollands, 2008, 2015; Marchesani, 2022; Marchesani et al., 2023). Ethical considerations in smart cities encompass a wide range of complex and interconnected issues that demand careful attention from managers and policymakers. Beyond technological aspects, these considerations extend to various ethical challenges associated with data collection, surveillance, and decision-making processes, inclusiveness and competitions in smart city projects.

The collection and use of personal data emerge as a prominent ethical concern within smart cities. The sheer volume of data generated by sensors, cameras, and mobile devices raises questions about privacy infringement. Individuals may express valid concerns regarding the gathering, storage, and sharing of their personal information without proper consent or knowledge. Striking a balance between data-driven innovation and protecting individual privacy becomes a crucial aspect of ethical smart city development. To address these concerns, it is imperative for managers and policymakers to prioritize the implementation of privacy-enhancing technologies and robust data security measures. Clear and comprehensive data governance frameworks should be established, outlining the principles and practices governing data collection, storage, access, and usage. Informed consent from individuals prior to data collection is essential to ensure transparency and respect for individual autonomy.

Another ethical consideration pertains to the potential misuse of collected data or unauthorized access to sensitive information. Safeguarding data against breaches, hacking, or exploitation becomes paramount to maintain public trust in smart city initiatives. Strong encryption methods, regular audits of data security measures, and protocols for reporting and addressing data breaches are vital steps in addressing these ethical concerns. Transparency and accountability form critical pillars of ethical smart city development. Managers and policymakers must provide citizens with clear and accessible information regarding the types of data being collected, the purposes for which it is being used, and the safeguards in place to protect it. Openness about data practices and decision-making processes fosters trust, encourages public engagement, and empowers citizens to hold decision-makers accountable.

Moreover, the use of algorithms in decision-making processes introduces ethical complexities within smart cities. Algorithms, if not carefully designed and audited, can inadvertently perpetuate biases, discrimination, and social inequalities. Biased algorithms can disproportionately impact marginalized communities and perpetuate existing social disparities. Regular audits of algorithms are necessary to identify and rectify any biases or discriminatory outcomes. Addressing the ethical dilemma of smart cities requires active engagement in public dialogue and consultation. Managers and policymakers should actively involve citizens, community organizations, and advocacy groups in shaping smart city policies. Incorporating diverse perspectives and ensuring representation from vulnerable and marginalized populations help to integrate ethical considerations into decision-making processes, resulting in more equitable and inclusive smart city initiatives.

Finally, the ethical considerations surrounding the smart city competition play a significant role as competition serves as a driving force for innovation and enables cities to stay competitive in the global landscape. With cities actively competing for talent, investments, and resources, competition fosters a culture of continuous improvement and encourages the development of innovative solutions to urban challenges. It motivates cities to explore new technologies, collaborate with various stakeholders, and adopt creative approaches to enhance livability, sustainability, and economic prosperity. Competition in smart city projects can take various forms, including competing for funding and grants, attracting private sector investments, or positioning themselves as leaders in specific domains such as mobility, energy efficiency, or digital governance. This competitive environment prompts cities to push boundaries, learn from one another, and leverage best practices to create smarter, more efficient, and sustainable urban environments. Nevertheless, striking a balance between inclusiveness and competition is of utmost importance in the context of smart city projects. Smart city initiatives need to be aware of the potential negative consequences that competition can have on social equity. The benefits of innovation should be shared by all residents, regardless of their background or socioeconomic status. This requires taking proactive measures to address any disparities and ensure equal access to resources and opportunities.

The rapid development of smart cities and the intense competition they often face can lead to imbalances in economic growth and livability. As cities strive to attract talent, investments, and resources, there is a risk of prioritizing economic interests at the expense of other essential aspects of urban life.

One concern is the increasing commodification of the city and its residents. As cities become more focused on economic competitiveness, there is a danger of viewing residents solely as productive assets or consumers rather than as individuals with diverse needs, aspirations, and rights. This can result in a dehumanizing effect, where the well-being and quality of life of residents are overshadowed by economic considerations. Another issue is the escalating housing prices and gentrification that often accompany the development of smart cities. As cities become more attractive and prosperous, the demand for housing rises, leading to skyrocketing prices and the displacement of lower income residents. This can disrupt established communities and exacerbate socioeconomic inequalities, as those with greater financial resources benefit while others struggle to find affordable housing options.

The pursuit of economic competitiveness can also lead to a focus on attracting high-skilled professionals, creating a potential brain drain from other regions or neglecting the needs of less advantaged communities. This concentration of talent and resources in certain areas can further deepen regional disparities and result in uneven development within the city. To address these challenges, it is crucial for smart city initiatives to adopt a balanced approach that considers both economic growth and the well-being of residents. This entails implementing policies that promote affordable housing options, support local businesses and entrepreneurship, and prioritize the social infrastructure necessary for inclusive and vibrant communities. It also requires safeguarding the rights and interests of residents, ensuring their active participation in decision-making processes, and mitigating the negative consequences of gentrification and economic disparity.

Thus, it assumes that while competition is a driving force for innovation and economic growth in smart cities, it is essential to recognize and address the potential imbalances and negative effects that can arise. Striking a balance between economic competitiveness and the well-being of residents is crucial to create sustainable and inclusive cities that prioritize the dignity and rights of all individuals. By taking a holistic approach that considers both economic and social dimensions, smart cities can foster equitable and livable urban environments for their residents.

8.4 Final Remark and Future Directions

In this book, we have discussed, praised, criticized, and contextualized the role of cities in contemporary society. This role, although not yet fully conceptualized, brings about significant changes in our current society, altering its dynamics, balances, and redefining its economic and social geography. Broadly speaking, the objective of smart cities is to improve urban performance by using data, digital technologies, and new organizational approaches to provide more efficient and sustainable services to citizens, monitor and optimize existing infrastructure, increase collaboration among different economic actors, and encourage innovative business models in both the private and public sectors. The outcomes highlighted, as well as the implementation and internal organization of smart cities, provide many ideas for future research and open up a myriad of assumptions and conjectures that will change over time with the advancement of technology and the digital readiness of users, leading to new and complex urban forms that are not yet fully predictable.

We assume that the adoption of smart city projects is a desirable dream for many cities. However, there is still a considerable distance between the city and its leading actors (i.e., urban managers, governance, citizens, entrepreneurs, companies, and external users). The way smart cities are organized and how these projects are implemented internally have not been extensively examined, and the actual and desired outcomes are not clearly evaluated due to the recent nature of this concept. It is assumed that smart cities are still at the beginning of their evolution and require a thorough theoretical, conceptual, and practical understanding.

Within this book, we have sought to systematically explore the re(shaping) of the city from an internal implementation and external outcomes perspective, building on international contributions in top-tier peer-reviewed journals, primary and secondary data, practical examples, and interviews with public managers, politicians, and city engineers to corroborate a concrete understanding of the current smart city. This book delves into the nature, dimensions, and dynamics of smart cities, deploying research and analysis from a wide variety of disciplines, including digital technologies, architecture, economic studies, regional studies, strategic management, and innovation studies.

Based on the comprehensive view highlighted in this book, interesting perspectives emerge pertaining to the evolutionary trajectory of cities in the future. Firstly, it is evident that cities will continue to evolve and adapt to technological advancements and changing social, economic, and environmental conditions. This necessitates the constant reevaluation and reimagining of urban planning and policy to ensure the resilience, sustainability, and inclusivity of cities. Secondly, the role of data, information technology, and digital tools in shaping urban environments will assume increasing significance. This calls for the development of new skills among urban professionals and citizens, as well as the creation of novel governance models and urban management strategies that effectively harness the potential of digital technologies. Lastly, the collaboration among diverse stakeholders, including governments, businesses, and citizens, will be pivotal for the successful implementation of smart city projects and the attainment of desired outcomes. This entails fostering a culture of innovation, openness, and communication, while also forging new partnerships and alliances.

It assumes that the exploration of these perspectives underscores the need for proactive and forward-thinking approaches in managing urban growth, ensuring that cities of the future are not only technologically advanced but also sustainable, inclusive, and responsive to the needs of their inhabitants. This assumption opens up a new vision of the smart city, highlighting the importance of noting that cities are undergoing a transformation, adopting a more business-like approach in their efforts to attract talented individuals who are sought-after resources in the realms of knowledge, economic capital, and human capital necessary for growth and competitiveness. Within this context, people are assuming pivotal roles as integral components in the functioning and progress of cities, transitioning from mere citizens to valuable contributors in the urban machinery.

In the global knowledge economy, cities that successfully attract and retain highly skilled professionals, scholars, entrepreneurs, and experts from diverse fields gain a significant competitive advantage. These individuals bring with them knowledge, expertise, and innovative ideas that fuel economic growth, job creation, and value generation for the local community. However, there is a risk of people being treated as mere "products" of cities, utilized as resources for development without genuine consideration for their well-being and personal fulfillment. To prevent cities from becoming mere economic machines that exploit human resources without offering satisfactory opportunities and living conditions, it is crucial to ensure a sustainable balance. Cities must invest in creating inclusive communities and prioritize the well-being of their residents. This entails promoting access to quality education, ensuring efficient public services, creating inclusive and sustainable urban spaces, and providing opportunities for civic participation and engagement.

Therefore, it is important to understand that the future of our cities lies in striking the right balance between leveraging technological advancements and human resources to create vibrant and sustainable urban environments while ensuring the well-being and personal fulfillment of citizens remain at the heart of urban planning and policy. As we navigate the complexities of the evolving urban landscape, let us remember that the true measure of a smart city lies not solely in its technological prowess or economic achievements, but in its ability to foster a thriving and inclusive community for all its residents.

Appendix: Methodological Approach

A1. Quantitative Approach

Sample of Analysis

The 30 Italian cities were selected on an equal basis within the territory to guarantee the high heterogeneity of the sample. To set the final sample, we based it on several constructs. First, we based on the homogeneity of the variables included in our model that has been considered uniform to assure the model approach. Second, we construct the sample of analysis based on the reference context as it is characterized by a different city size with a population ranging from 67,200 to 285,200. Finally, to guarantee the robustness of our sample, we apply the probability-proportional-to-size sampling in each stratum of the medium and large cities. This distinction between large and medium size has also been used as a control variable in our model.

Using this approach, the city with the largest population in each stratum emerges as a representative sample to be considered (Levy & Lemeshow, 2011). As a result, the final sample of analysis considered 30 of the 46 most populated Italian cities to guarantee the heterogeneity of the analysis sample. Thus, we use geo-software to draw the Italian map with our sample of analysis emerging from the sampling based on average from 2010 to 2021 (See Fig. A1), which shows the distribution and size of the cities under consideration. This empirical investigation focus on the Italian context as it currently is at the center of the academic and political debate on firms' green transformation, social policies and the current trends in sustainable advance, constituting a prominent empirical focus for the debates on sustainable development in Europe compared to other countries (i.e., Asia, US and Latin America).

Estimation Model

Due to the intrinsic characteristics of the panel dataset, which consists of a wide cross-section and a relatively short time period of 11 years, the dynamic Generalized Methods of Moments (GMM) estimation technique was employed in this article. By including time and city-fixed effects, the time-varying nature of the data was accounted for, and unobserved

Fig. A1. Sample of Analysis. *Source*: Created by author.

heterogeneity across cities was controlled. Following Roodman's approach (Roodman, 2009a, 2009b, 2009c) to the estimation model, internal instruments from lagged variables were also utilized to address the issue of over-identification (Windmeijer, 2005).

Roodman's approach (Roodman, 2009a, 2009b, 2009c) was employed in our estimation model to address the critical issue of identification and ensure the validity of our results. This approach relies on the use of internal instruments derived from lagged variables, and it was chosen for several reasons. Firstly, identification is a crucial aspect in estimating econometric models as it allows for the unique attribution of the effects of independent variables on the dependent variable. However, in the presence of endogenous or correlated variables, identification can become challenging. The use of internal instruments, such as lagged variables, helps overcome this issue by providing additional information that aids in the identification process. Roodman's approach offers a robust solution for addressing over-identification, a common concern in econometric analysis. Over-identification occurs when there are more instruments than necessary to identify the parameters of interest, leading to inefficient estimates and potential biases. By employing internal instruments from lagged variables, we ensure that the model is neither under-identified nor over-identified, striking a balance that promotes accurate and reliable estimation (Wooldridge, 2002). The inclusion of lagged variables as instruments is particularly valuable in dynamic models, where the dependent variable's past values can serve as relevant predictors. By capturing the persistence and lagged effects of the dependent variable, the lagged variables act as valid instruments, allowing us to account for the dynamic nature of the relationship under investigation. Furthermore, Roodman's approach is well-suited for panel datasets with a wide cross-section and a relatively short time period, as is the case in our study (Roodman, 2009c). By incorporating time and city-fixed effects, we account for the time-varying characteristics of the data and control for unobserved heterogeneity across cities This helps mitigate potential biases and provides a more accurate estimation of the effects of interest.

To ensure the validity of the dynamic GMM estimations and assess the robustness of the results, various diagnostic tests were conducted. Specifically, the Arellano and Bond test, the Sargan test, and the Wald Chi-Square test were employed (Bond & Arellano, 2012; Magazzini & Calzolari, 2020; Roodman, 2009b; Sargan, 1958).

The Sargan test was used to examine the joint validity of the instruments and detect over-identification in the model (Sargan, 1958). The choice of using the Sargan test instead of the Hausmann test was made, as it is known to be more effective in detecting over-identification and less susceptible to the instrument proliferation problem, as noted by Roodman (Roodman, 2009c). Specifically, unlike the Hausmann test, the Sargan test is known to be more effective in detecting over-identification and less susceptible to the instrument proliferation problem (Bond & Arellano, 2012; Breitung & Salish, 2021; Debarsy, 2012). By conducting the Sargan test, we ensured that our instruments were not correlated with the error term and that the model was properly identified. This test further enhanced the validity and reliability of our estimation results, contributing to the robustness of our analysis.

In addition, the Arellano and Bond test (Bond & Arellano, 2012; Lee & Yu, 2014) was applied to assess any autocorrelation in the idiosyncratic disturbance term. Further investigation of potential autocorrelation issues was conducted through the AR(1) and AR(2) tests. These tests helped evaluate the presence of autocorrelation and address it appropriately.

The Arellano and Bond test is a widely used method to detect autocorrelation in dynamic panel data models. It is based on the first-differenced transformation of the model equation and compares the coefficient estimates of lagged dependent variables with their predicted values. The test examines whether the lagged dependent variables are correlated with the first-differenced residuals, indicating the presence of autocorrelation. To further investigate potential autocorrelation issues, we conducted the AR(1) and AR(2) tests. The AR(1) test examines first-order autocorrelation by regressing the residuals on lagged residuals from the previous time period. Similarly, the AR(2) test examines second-order autocorrelation by regressing the residuals on two lags of residuals. These tests help us evaluate the presence and magnitude of autocorrelation in the error terms. If significant autocorrelation is detected, it suggests that the error terms are correlated over time, violating the assumption of independence. To address autocorrelation, we can employ various techniques such as adding lagged dependent variables, including additional lagged independent variables, or using alternative estimation methods such as Generalized Method of Moments (GMM) or Fixed Effects (FE) models (Allison, 1994; Elhorst, 2003; Wooldridge, 2002). By applying the Arellano and Bond test, along with the AR(1) and AR(2) tests, we can identify and address autocorrelation in our model. This ensures that our estimations account for the time-dependent patterns and improves the reliability of our results. By taking these rigorous steps to detect and mitigate autocorrelation, we enhance the validity and robustness of our analysis, providing more accurate insights into the relationship between variables in our dynamic panel data model.

Finally, to ensure the validity of the model and address heteroskedasticity, the Wald Chi-square test was conducted along with additional checks. Heteroskedasticity refers to the presence of unequal variances in the error terms, which can affect the efficiency and accuracy of the estimations (Lee & Yu, 2014; Roodman, 2009c). The Wald Chi-square test is a statistical test that examines whether the coefficients in the model are jointly significant. It assesses the overall validity of the model and helps determine if the estimated relationships between variables are statistically meaningful. To mitigate the effects of unobserved heterogeneity and endogeneity issues, the Roodman GMM dynamic approach (Roodman, 2009c) was adopted. This approach is a robust estimation method that addresses endogeneity and unobserved heterogeneity by using instrumental variables. In this case, the lagged-dependent variable was utilized as an instrument to capture the persistence of the dependent variable. By including the lagged-dependent variable as an instrument, potential biases caused by the endogeneity of the dependent variable and the presence of unobserved factors that may affect the relationship between variables are controlled. By employing the Wald Chi-square test, implementing additional checks, and adopting the Roodman GMM dynamic approach, the validity and reliability of the model are enhanced. These methodological approach enable the control of heteroskedasticity, address endogeneity and unobserved heterogeneity, and provide more accurate and robust estimations (Lee & Yu, 2014; Magazzini & Calzolari, 2020; Roodman, 2009a). As a result, meaningful conclusions can be drawn, and reliable inferences about the relationships and dynamics within the model can be made.

As a last check, the presence of multicollinearity, which refers to high correlation among independent variables in a regression model, was assessed using the Variance Inflation Factor (VIF). The VIF measures the extent to which the variance of an estimated regression coefficient is inflated due to multicollinearity (Akinwande, Dikko, & Samson, 2015). In our analysis, a VIF value of less than four is generally considered acceptable, indicating that the independent variables in the model are not highly correlated with each other. This suggests that multicollinearity is not a significant issue in our model. By ensuring low levels of multicollinearity, we enhance the stability and reliability of the results. High levels of multicollinearity can lead to unreliable coefficient estimates, making it difficult to interpret the impact of individual variables on the dependent variable. However, in our case, the VIF values below four indicate that the variables in our model are relatively independent and do not suffer from multicollinearity issues. Having a stable and reliable model allows us to make accurate inferences and draw meaningful conclusions from our analysis. It ensures that the estimated coefficients are more robust and the relationships between variables are accurately represented. Therefore, by assessing and addressing multicollinearity, we can have greater confidence in the validity and accuracy of our results.

By employing these rigorous statistical techniques and diagnostic tests, the robustness of the analysis was enhanced, and potential biases were minimized. These methodological choices strengthen the validity of the findings and provide a solid foundation for drawing meaningful conclusions from the research.

Data Collection and Variables

This research uses city-level variables and collects data from databases at the national (ANPR, AGCOM, ISTAT, GreenItaly, GSE, OECD, MIUR CoC and IBS) and international (EUROSTAT and OECD) levels. Specifically, in Tables A1 and A2, describe the variables adopted in the empirical analysis showing the source, type of variables, operationalization, and short description. Table A3 displays the results of the GMM model.

Table A1. Variables Description and Construction.

Variable	Description	Operationalization	Type	Source
Employment	Percentage rate of employment in the city (over 30 years)	Natural logarithm of the variable over the population yearly	Constant	ISTAT
Business creation	Number of companies registered in the "Chamber of Commerce" of each city per year	Natural logarithm of the variable over the total number of companies yearly	Constant	ISTAT, CoC
Start-ups	Number of companies registered in the "Start-Up" section in the "Chamber of Commerce" of each city per year	Natural logarithm of the variable over the total number of companies yearly	Constant	ISTAT, CoC
Innovative companies	Number of innovative companies registered in the "Chamber of Commerce" of each city per year	Natural logarithm of the variable over the total number of companies yearly	Constant	ISTAT, CoC
Female entrepreneurship	The total number of companies founded by female entrepreneurs in the city per year	Natural logarithm of the variable over the total number of companies yearly	Constant	ISTAT, CoC
Green companies	Companies with ISO 9001, ISO 14001, and EMAS (Eco-Management and Audit Scheme) certifications founded yearly in the city	Natural logarithm of the variable over the total companies in the city yearly	Constant	ISTAT, GreenItaly
SC digital implementation	Variable constructed based on nine indicators relevant to the technological development of the city (refer to Table A2)	Index [0-1] based on the indicators, populations, and distribution in the sample	Index	Multiple sources
SC amenities and facilities	Variable constructed based on seven indicators relevant to the liveability of the smart city (refer to Table A2)	Index [0-1] based on the indicators, populations, and distribution in the sample	Index	Multiple sources
City development	Economic development of the city according to the division proposed by the European community	Dummy variable that considers cities in the most developed urban areas (1) and cities in transition areas (0)	Dummy	ISTAT

Table A1. *(Continued)*

Variable	Description	Operationalization	Type	Source
City size	Size of the city considering that 300,000 is the threshold between medium and large cities	Dummy variable considers (1) cities that have a population greater than 300,000 and (0) cities that have a lower population	Dummy	ISTAT
Employment	Percentage rate of employment in the city (over 30 years)	Natural logarithm of the variable over the population yearly	Constant	ISTAT
City GDP	Gross domestic product produced in each city in the year *n* considered	Natural logarithm of the variable over the population in the city	Constant	EUROSTAT – OECD
Population	Descriptive number of the total population in the city in the year "*n*"	Natural logarithm of the variable	Constant	ISTAT-ANPR
Airports	Total number of airport (public and private) in the city and within 50 km	Natural logarithm of the variable over the population in the city	Constant	ISTAT
Private R&D	Total amount of private sector investment in research & development in the city per year	Natural logarithm of the variable over the population in the city	Constant	ISTAT
Public R&D	Total amount of public sector investment in research & development in the city per year	Natural logarithm of the variable over the population in the city	Constant	ISTAT
Total companies	Total number of firms active in the city, based on the registration on the Chamber of Commerce	Natural logarithm of the variable over the population in the city	Constant	IBS

Source: Table created by author.

ANPR (Anagrafe Nazionale della Popolazione Residente); ISTAT (Italian National Institute of Statistics); GSE (Energetic Services Management); GreenItaly; Terna (Italian Energetic Network); OECD (Organisation for Economic Co-operation and Development); IBS (Italian Business Register); EUROSTAT (Statistical office of the European Union).

Table A2. Independent Variables Construction.

Variable	Description	Operationalization	Type	Source
Smart City – Digital Technology Implementation				
Online services	Total number of online services proposed by the city on the municipal website per year	Natural logarithm of the variable	Constant	ISTAT
Broadband access	Percentage of families who have access to ADSL	Natural logarithm of the variable operationalized over the population in the city per year	Constant	AGCOM
Municipal app	Municipal app download number per year	Natural logarithm of the variable over the population in the city	Constant	ISTAT
Home-banking diffusion	Number of users who utilize home banking in the city per year	Natural logarithm of the variable over the population in the city	Percentage	ISTAT
Digital transparency	Number of public data concerning the investments of the city per year	Natural logarithm of the variable over the population in the city	Constant	ANAC
Digital openness	Total number of public access databases in the city per year	Natural logarithm of the variable over the population in the city	Constant	FPA
Social public administration	Variable based on engagement, productivity, and use of public online services in the city per year	Natural logarithm of the variable over the population in the city	Construct	FPA
Public Wi-Fi	Variable based on the number of access points, quality of service, and communication in the city per year	Natural logarithm of the variable over the population in the city	Construct	ISTAT
IoT development	The total amount of investments in IoT and ICTs in the city per year	Natural logarithm of the variable over the population in the city	Percentage	ISTAT
Smart City – Amenities and Facilities				
Health care	Total number of public and private hospitalization services in the city per year	Natural logarithm of the variable over the population in the city	Constant	ISTAT
Health personnel	Total number of employees in public and private hospitals in the city per year	Natural logarithm of the variable over the population in the city	Constant	ISTAT-IPS

Table A2. *(Continued)*

Variable	Description	Operationalization	Type	Source
Elderly assistance	Total number of public and private elderly reception services in the city per year	Natural logarithm of the variable over the population in the city	Constant	ISTAT-IPS
Hospital emigration	Total number of hospital emigrations to other regions for ordinary hospitalizations in the city per year	Natural logarithm of the variable over the population in the city	Constant	ISTAT-IPS
Childcare	Total number of childcare assistance based on private and public services in the city per year	Natural logarithm of the variable over the population in the city	Constant	ISTAT
Job integration	Total number of new residents who work permanently in the city and new residents who found a job in the first six months in the city per year	Natural logarithm of the variable over the population in the city	Percentage	ISTAT
Coworking area	Total number of coworking services in the city per year	Natural logarithm of the variable over the total number of companies in the city	Percentage	FPA-OECD
Education development	Number of people in the city who interacted with services offered in the educational field (public schools, courses, master's, or services in the educational field) in the city per year	Natural logarithm of the variable over the population in the city	Percentage	MIUR

Source: Table created by author.

AGCOM (Autorità per le Garanzie nelle Comunicazioni); ANAC (Autorità Nazionale Anticorruzione); FPA (Innovation in Public Administration); ISTAT (Italian National Institute of Statistics); IPS (Intrusion Prevention System); OECD (Organization for Economic Co-operation and Development); MIUR (Ministero dell'Istruzione e del Merito).

Table A3. GMM Model Estimation Results – The Effect of SC Implementation on Economic Outcomes.

	Model I		Model II		Model III		Model IV		Model V		Model VI	
	Coeff.	**s.e.**	**Coeff.**	**s.e.**	**Coeff.**	**s.e.**	**Coeff.**	**s.e.**	**Coeff.**	**s.e.**	**Coeff.**	**s.e.**
Employment ($t-1$)	0.492***	[0.112]										
Business creation ($t-1$)			0.573***	[0.083]								
Start-ups ($t-1$)					0.668***	[0.096]						
Innovative companies ($t-1$)							0.701***	[0.106]				
Green companies ($t-1$)									0.554**	[0.168]		
Female entrepreneurship ($t-1$)											0.648***	[0.120]
SC digital implementation	0.332**	[0.243]	0.492***	[0.112]	0.613***	[0.105]	0.698***	[0.123]	0.597***	[0.096]	0.736**	[0.108]
SC amenities and facilities	0.468***	[0.612]	0.265***	[0.232]	0.258**	[0.216]	0.227**	[0.183]	0.558***	[0.072]	0.363***	[0.334]
City development	0.106**	[0.212]	0.188***	[0.212]	0.132**	[0.268]	0.144***	[0.283]	0.141*	[0.261]	0.176**	[0.294]
City size	0.271*	[0.442]	0.371**	[0.442]	0.187**	[0.431]	0.199*	[0.466]	0.196**	[0.382]	0.231*	[0.354]
Employment	–	–	0.451*	[0.112]	0.543	[0.331]	0.332	[0.293]	0.198*	[0.301]	0.267*	[0.288]
City GDP	0.216**	[0.185]	0.199**	[0.185]	0.214**	[0.198]	0.226**	[0.213]	0.223*	[0.188]	0.258*	[0.258]
Population	0.324*	[0.188]	0.406**	[0.188]	0.319*	[0.204]	0.331**	[0.243]	0.328*	[0.218]	0.363**	[0.233]
Airports	1.366*	[0.236]	1.107*	[0.236]	1.064	[0.245]	1.076*	[0.230]	1.073***	[0.216]	1.108***	[0.281]
Private R&D	−0.153**	[0.271]	−0.199*	[0.271]	−0.113***	[0.255]	−0.101**	[0.271]	−0.104*	[0.234]	−0.069**	[0.260]
Public R&D	0.309**	[0.138]	0.178***	[0.138]	0.267*	[0.144]	0.255*	[0.152]	0.258**	[0.128]	0.223*	[0.150]
Total companies	1.072*	[0.402]	1.352***	[0.402]	0.853**	[0.578]	0.841***	[0.618]	0.844**	[0.539]	0.809**	[0.701]
City effect	*Included*		*Included*		*Included*		*Included*		*Included*		*Included*	
Year effect	*Included*		*Included*		*Included*		*Included*		*Included*		*Included*	

Wald chi²	365.32	401.23	386.43	379.22	392.67	361.81
AR(2)	1.87	1.34	2.26	2.08	1.95	2.13
p Value	0.113	0.127	0.168	0.142	0.117	0.144
Sargan test	46.35	48.24	59.16	53.18	51.93	64.85
p Value	0.669	0.745	0.692	0.633	0.659	0.704
No observation	330	330	330	330	330	330
No* city	30	30	30	30	30	30

Source: Table created by author.

Employment, business creation, start-ups, innovative companies, green companies and female entrepreneurship as a DV. Significance level: $*p < 0.05$; $**p < 0.01$; $***p < 0.001$. Robust standard errors in brackets. VIF (variance inflation factor) < 4. Arellano–Bond AR (2) is used to look for possible autocorrelation issues. Sargan test is used to look for possible over-identification restrictions in the model. The variable "Employment" as a control is not considered in Model 1 as it is used as the DV. The number of observations is reduced (360–330) for the leg effect (−1).

References

Akinwande, M. O., Dikko, H. G., & Samson, A. (2015). Variance inflation factor: As a condition for the inclusion of suppressor variable(s) in regression analysis. *Open Journal of Statistics, 05*(07). doi:10.4236/ojs.2015.57075

Allison, P. D. (1994). Using panel data to estimate the effects of events. *Sociological Methods & Research*. doi:10.1177/0049124194023002002

Arellano, M., & Bond, S. (1991). Some tests of specification for panel data: Monte Carlo evidence and an application to employment equations. *The Review of Economic Studies, 58*.

Bond, S., & Arellano, M. (2012). Some tests of specification for panel data: Monte Carlo evidence and an application to employment equations. *The Review of Economic Studies, 58*(2).

Breitung, J., & Salish, N. (2021). Estimation of heterogeneous panels with systematic slope variations. *Journal of Econometrics, 220*(2). doi:10.1016/j.jeconom.2020.04.007

Debarsy, N. (2012). The Mundlak approach in the spatial durbin panel data model. *Spatial Economic Analysis, 7*(1). doi:10.1080/17421772.2011.647059

Elhorst, J. P. (2003). Specification and estimation of spatial panel data models. *International Regional Science Review, 26*(3). doi:10.1177/0160017603253791

Lee, L. F., & Yu, J. (2014). Efficient GMM estimation of spatial dynamic panel data models with fixed effects. *Journal of Econometrics, 180*(2). doi:10.1016/j.jeconom.2014.03.003

Magazzini, L., & Calzolari, G. (2020). Testing initial conditions in dynamic panel data models. *Econometric Reviews, 39*(2). doi:10.1080/07474938.2019.1690194

Roodman, D. (2009a). A Note on the theme of too many instruments. *Oxford Bulletin of Economics & Statistics, 71*(1). doi:10.1111/j.1468-0084.2008.00542.x

Roodman, D. (2009b). How to do xtabond2: An introduction to difference and system GMM in Stata. *STATA Journal, 9*(1). doi:10.1177/1536867x0900900106

Roodman, D. (2009c). Practitioners' corner: A note on the theme of too many instruments. *Oxford Bulletin of Economics & Statistics, 71*(1). doi:10.1111/j.1468-0084.2008.00542.x

Sargan, J. D. (1958). The estimation of economic relationships using instrumental variables. *Econometrica, 26*(3). doi:10.2307/1907619

Windmeijer, F. (2005). A finite sample correction for the variance of linear efficient two-step GMM estimators. *Journal of Econometrics, 126*(1), 25–51. doi:10.1016/j.jeconom.2004.02.005

Wooldridge, J. M. (2002). *Econometric analysis of cross section and panel data*. Booksgooglecom (Vol. 58(2)). doi:10.1515/humr.2003.021

A2. Interview Appendix

To collect data on the internal implementation of the smart city ecosystem (part one of the book), a series of interviews were conducted with public managers and policymakers responsible for various functions and areas within the municipalities. These interviews aimed to explore the implementation of smart city projects, the capabilities involved, key actors, and the connections between the internal environment and institutions in the development of these projects.

The interviews took place between January and May 2023 and involved nine participants from different municipal departments, including the Digitalization office, tourism, mobility, control room, policies, urban investments, and front offices. The respondents held diverse job titles and had responsibilities and competencies relevant to their respective roles. During the interviews, participants were asked a range of questions to delve into the specific aspects of the digitalization projects in their cities. The questions covered topics such as the functioning and phases of the selected project, the activities carried out within each phase, the benefits observed from its implementation, and the challenges encountered. Additionally, the interviews explored the technological aspects of the projects, including the specific technologies used in different phases, how the participants became acquainted with these technologies, and whether they were sourced internally or through collaborations with external actors. The types of data involved in the projects and their collection and processing methods were also discussed. Furthermore, the interviews delved into the soft skills and competencies of the actors involved in the projects. Participants were asked about the actors' roles, their specific competencies, how contact was established with them, and whether the required professional skills were available within the organization or obtained through external collaborations.

The final set of questions focused on the project's evaluation, control, and monitoring. This included inquiries about how the benefits generated from the implementation were identified (ex-ante), how the results were evaluated (ex-post), who was responsible for the project's functioning, and whether the project had a positive impact on the local ecosystem.

Overall, these interviews provided valuable insights into the implementation of smart city projects, the coordination of different dimensions, and the role of various stakeholders. The information collected through these interviews contributes to a deeper understanding of the factors influencing the success and challenges of smart city initiatives.

The complete list of interviewees can be found in Table B1.

Table B1. List of Interviews.

Interview No.	Date	Time	Length (minutes)	Role within the Municipality
1	17-Gen	15:00	46:00	Digital transition sector manager
2	09-Feb	09:30	65:00	Manager of digital service and telematics implementation
3	09-Mar	10:00	38:00	Corruption prevention office manager
4	15-Mar	10:00	45:00	Registry and demographic services manager
5	15-Mar	11:00	50:00	Civil status office manager
6	04-Apr	10:30	52:00	Technical project management office for european projects
7	04-Apr	11:15	30:00	Monitoring and reporting office
8	22-Mag	15:30	64:00	Manager smart mobility
9	23-Mag	09:00	22:00	Manager control room and security

Source: Table created by author.

Exploratory interviews: Municipalities β.

References

Abid, N., Marchesani, F., Ceci, F., & Masciarelli, F. (2023). Assessing capabilities to embrace digital transformation: The case of Southern Italy. *Lecture Notes in Information Systems and Organisation*, *59*, 169–182. doi:10.1007/978-3-031-15770-7_11

Abid, N., Marchesani, F., Ceci, F., Masciarelli, F., & Ahmad, F. (2022). Cities trajectories in the digital era: Exploring the impact of technological advancement and institutional quality on environmental and social sustainability. *Journal of Cleaner Production*, *377*(September), 134378. doi:10.1016/j.jclepro.2022.134378

Adler, P., & Florida, R. (2021). The rise of urban tech: How innovations for cities come from cities. *Regional Studies*, *55*(10–11), 1787–1800. doi:10.1080/00343404.2021.1962520

Adler, P., Florida, R., King, K., & Mellander, C. (2019). The city and high-tech startups: The spatial organization of Schumpeterian entrepreneurship. *Cities*, *87*(January), 121–130. doi:10.1016/j.cities.2018.12.013

Adner, R., & Kapoor, R. (2010). Value creation in innovation ecosystems: How the structure of technological interdependence affects firm performance in new technology generations. *Strategic Management Journal*, *31*(3). doi:10.1002/smj.821

Akinwande, M. O., Dikko, H. G., & Samson, A. (2015). Variance inflation factor: As a condition for the inclusion of suppressor variable(s) in regression analysis. *Open Journal of Statistics*, *05*(07). doi:10.4236/ojs.2015.57075

Albino, V., Berardi, U., & Dangelico, R. M. (2015). Smart cities: Definitions, dimensions, performance, and initiatives. *Journal of Urban Technology*. doi:10.1080/10630732.2014.942092

Albuquerque, V., Oliveira, A., Barbosa, J. L., Rodrigues, R. S., Andrade, F., Dias, M. S., & Ferreira, J. C. (2021). Smart cities: Data-driven solutions to understand disruptive problems in transportation—The Lisbon case study. *Energies*, *14*(11). doi:10.3390/en14113044

Aletà, N. B., Alonso, C. M., & Ruiz, R. M. A. (2017). Smart Mobility and Smart Environment in the Spanish cities. *Transportation Research Procedia*, *24*, 163–170. doi:10.1016/j.trpro.2017.05.084

Allam, Z., & Dhunny, Z. A. (2019). On big data, artificial intelligence and smart cities. *Cities*, *89*(December 2018), 80–91. doi:10.1016/j.cities.2019.01.032

Allam, Z., & Jones, D. S. (2020). Pandemic stricken cities on lockdown. Where are our planning and design professionals [now, then and into the future]? *Land Use Policy*. doi:10.1016/j.landusepol.2020.104805

Allen, A., & Bartlett, T. (2022). Navigating stigma through everyday city-making: Gendered trajectories, politics and outcomes in the periphery of Lima. *Urban Studies Journal Limited*, *59*(3), 2021. doi:10.1177/00420980211044409

Allison, P. D. (1994). Using panel data to estimate the effects of events. *Sociological Methods & Research*. doi:10.1177/0049124194023002002

Almirall, E., Wareham, J., Ratti, C., Conesa, P., Bria, F., Gaviria, A., & Edmondson, A. (2016). Smart Cities at the crossroads: New tensions in city transformation. *California Management Review*, *59*(1), 141–152. doi:10.1177/0008125616683949

Angelidou, M., Politis, C., Panori, A., Barkratsas, T., & Fellnhofer, K. (2022). Emerging smart city, transport and energy trends in urban settings: Results of a pan-European foresight exercise with 120 experts. *Technological Forecasting and Social Change*, *183*(July), 121915. doi:10.1016/j.techfore.2022.121915

Anttila, J., & Jussila, K. (2018). Universities and smart cities: The challenges to high quality. *Total Quality Management and Business Excellence*, *29*(9–10), 1058–1073. doi:10.1080/14783363.2018.1486552

Appio, F. P., Lima, M., & Paroutis, S. (2019). Understanding Smart Cities: Innovation ecosystems, technological advancements, and societal challenges. *Technological Forecasting and Social Change*, *142*(December 2018), 1–14. doi:10.1016/j.techfore.2018.12.018

Arabindoo, P. (2020). Renewable energy, sustainability paradox and the post-urban question. *Urban Studies*, *57*(11), 2300–2320. doi:10.1177/0042098019885080

Arbolino, R., & Boffardi, R. (2017). The impact of institutional quality and efficient cohesion investments on economic growth evidence from Italian regions. *Sustainability (Switzerland)*, *9*(8). doi:10.3390/su9081432

Archibugi, D., Filippetti, A., & Frenz, M. (2020). Investment in innovation for European recovery: A public policy priority. *Science and Public Policy*, *47*(1), 92–102. doi:10.1093/scipol/scz049

Arellano, M., & Bond, S. (1991). Some tests of specification for panel data: Monte Carlo evidence and an application to employment equations. *The Review of Economic Studies*, *58*.

Asthana, S., & Dwivedi, A. (2020). Performance measurement of India-based third party logistics sector: An empirical study of user versus provider perspectives. *Production Planning & Control*, *31*(2–3), 259–272. doi:10.1080/09537287.2019.1631467

Bacq, S., & Janssen, F. (2011). The multiple faces of social entrepreneurship: A review of definitional issues based on geographical and thematic criteria. *Entrepreneurship & Regional Development*. doi:10.1080/08985626.2011.577242

Badii, C., Bellini, P., Cenni, D., Difino, A., Nesi, P., & Paolucci, M. (2017). Analysis and assessment of a knowledge based smart city architecture providing service APIs. *Future Generation Computer Systems*, *75*. doi:10.1016/j.future.2017.05.001

Bakici, T., Almirall, E., & Wareham, J. (2013). A smart city initiative: The case of Barcelona. *Journal of the Knowledge Economy*, *4*(2), 135–148. doi:10.1007/s13132-012-0084-9

Bania, N., Eberts, R. W., & Fogarty, M. S. (1993). Universities and the startup of new companies: Can we generalize from Route 128 and Silicon Valley? *The Review of Economics and Statistics*. doi:10.2307/2110037

Barbi, M., & Mattioli, S. (2019). Human capital, investor trust, and equity crowdfunding. *Research in International Business and Finance*. doi:10.1016/j.ribaf.2019.02.005

Barham, H., & Daim, T. (2020). The use of readiness assessment for big data projects. *Sustainable Cities and Society*, *60*(November 2019), 102233. doi:10.1016/j.scs.2020.102233

Bartoloni, S., Calò, E., Marinelli, L., Pascucci, F., Dezi, L., Carayannis, E., … Gregori, G. L. (2022). Towards designing society 5.0 solutions: The new Quintuple Helix – Design Thinking approach to technology. *Technovation*, *113*(November 2021). doi:10.1016/j.technovation.2021.102413

Bastidas-Manzano, A. B., Sánchez-Fernández, J., & Casado-Aranda, L. A. (2021). The past, present, and future of smart tourism destinations: A bibliometric analysis. *Journal of Hospitality & Tourism Research*, *45*(3). doi:10.1177/1096348020967062

Batabyal, A. A., & Nijkamp, P. (2019). Creative capital, information and communication technologies, and economic growth in smart cities. *Economics of Innovation and New Technology*, *28*(2). doi:10.1080/10438599.2018.1433587

Battarra, R., Gargiulo, C., Tremiterra, M. R., & Zucaro, F. (2018). Smart mobility in Italian metropolitan cities: A comparative analysis through indicators and actions. *Sustainable Cities and Society*, *41*(July 2017), 556–567. doi:10.1016/j.scs.2018.06.006

Batty, M., Axhausen, K. W., Giannotti, F., Pozdnoukhov, A., Bazzani, A., Wachowicz, M., … Portugali, Y. (2012). Smart cities of the future. *European Physical Journal: Special Topics*, *214*(1), 481–518. doi:10.1140/epjst/e2012-01703-3

Baykal, A. (2012). Green roofs Copenhagen. In D. Rømø (Ed.), *The technical and environmental administration in city of Copenhagen*; Københavns Kommune. Retrieved from https://www.klimatilpasning.dk/media/1810105/green-roofs-copenhagen.pdf

Benevolo, C., Dameri, R. P., & Auria, B. D. (2016). *Empowering organizations: Enabling platforms and artefacts* (Vol. 11, p. 315). doi:10.1007/978-3-319-23784-8

Benevolo, C., Dameri, R. P., & D'Auria, B. (2016). *Smart Mobility in Smart City*. doi:10.1007/978-3-319-23784-8_2

Bernardi, M., & Diamantini, D. (2018). Shaping the sharing city: An exploratory study on Seoul and Milan. *Journal of Cleaner Production*, *203*. doi:10.1016/j.jclepro.2018.08.132

Berrone, P., Ricart, J. E., & Carrasco, C. (2016). The open kimono: Toward a general framework for open data initiatives in cities. *California Management Review*, *59*(1), 39–70. doi:10.1177/0008125616683703

Betz, M. R., Partridge, M. D., & Fallah, B. (2016). Smart cities and attracting knowledge workers: Which cities attract highly-educated workers in the 21st century? *Papers in Regional Science*, *95*(4), 819–841. doi:10.1111/pirs.12163

Beugelsdijk, S., Kostova, T., & Roth, K. (2017). An overview of Hofstede-inspired country-level culture research in international business since 2006. *Journal of International Business Studies*, *48*. doi:10.1057/s41267-016-0038-8

Bibri, S. E., & Krogstie, J. (2020). The emerging data–driven Smart City and its innovative applied solutions for sustainability: The cases of London and Barcelona. *Energy Informatics*, *3*(1). doi:10.1186/s42162-020-00108-6

Bikfalvi, A., Jäger, A., & Lay, G. (2014). The incidence and diffusion of teamwork in manufacturing – Evidences from a Pan-European survey. *Journal of Organizational Change Management*, *27*(2), 206–231. doi:10.1108/JOCM-04-2013-0052

Black, D., & Henderson, V. (1999). A theory of urban growth. *Journal of Political Economy*, *107*(2). doi:10.1086/250060

Bocarejo, J. P., Portilla, I. J., Velásquez, J. M., Cruz, M. N., Peña, A., & Oviedo, D. R. (2014). An innovative transit system and its impact on low income users: The case of the Metrocable in Medellín. *Journal of Transport Geography, 39*. doi:10.1016/j.jtrangeo.2014.06.018

Boes, K., Buhalis, D., & Inversini, A. (2015). Conceptualising smart tourism destination dimensions. In *Information and Communication Technologies in Tourism 2015*. doi:10.1007/978-3-319-14343-9_29

Bonakdar, A., & Audirac, I. (n.d.). City branding and the link to urban planning: Theories, practices, and challenges. *Journal of Planning Literature, 2020*(2), 147–160. doi:10.1177/0885412219878879

Bond, S., & Arellano, M. (2012). Some tests of specification for panel data: Monte Carlo evidence and an application to employment equations. *The Review of Economic Studies, 58*(2), 277–297

Borrás, S., & Edler, J. (2020). The roles of the state in the governance of socio-technical systems' transformation. *Research Policy, 49*(5), 103971. doi:10.1016/j.respol.2020.103971

Boschma, R. A. (2005). Does geographical proximity favour innovation? *Économie et Institutions*. (6–7), 111–128. doi:10.4000/ei.926

Bourliataux-Lajoinie, S., Dosquet, F., & del Olmo Arriaga, J. L. (2019). The dark side of digital technology to overtourism: The case of Barcelona. *Worldwide Hospitality and Tourism Themes, 11*(5). doi:10.1108/WHATT-06-2019-0041

Breitung, J., & Salish, N. (2021). Estimation of heterogeneous panels with systematic slope variations. *Journal of Econometrics, 220*(2). doi:10.1016/j.jeconom.2020.04.007

Breslin, S., & Wilson, J. D. (2015). Towards Asian regional functional futures: Bringing Mitrany back in? *Australian Journal of International Affairs*. doi:10.1080/10357718.2014.978736

Breslow, H. (2021). The smart city and the containment of informality: The case of Dubai. *Urban Studies, 58*(3), 471–486. doi:10.1177/0042098020903233

Buhalis, D. (2020). Technology in tourism-from information communication technologies to eTourism and smart tourism towards ambient intelligence tourism: A perspective article. *Tourism Review, 75*(1), 267–272. doi:10.1108/TR-06-2019-0258

Buhalis, D., & Amaranggana, A. (2015). Smart tourism destinations enhancing tourism experience through personalisation of services. In *Information and Communication Technologies in Tourism 2015*. doi:10.1007/978-3-319-14343-9_28

Buhalis, D., O'Connor, P., & Leung, R. (2023). Smart hospitality: From smart cities and smart tourism towards agile business ecosystems in networked destinations. *International Journal of Contemporary Hospitality Management, 35*(1), 369–393. doi:10.1108/IJCHM-04-2022-0497

Buhalis, D., & Wagner, R. (2013). E-destinations: Global best practice in tourism technologies and applications. In *Information and Communication Technologies in Tourism, 2013* (Mintel 2011, pp. 119–130). doi:10.1007/978-3-642-36309-2_11

Burns, R., Fast, V., Levenda, A., & Miller, B. (2021). Smart cities: Between worlding and provincialising. *Urban Studies, 58*(3), 461–470. doi:10.1177/0042098020975982

Camboim, G. F., Zawislak, P. A., & Pufal, N. A. (2019). Driving elements to make cities smarter: Evidences from European projects. *Technological Forecasting and Social Change, 142*(December 2017), 154–167. doi:10.1016/j.techfore.2018.09.014

Capdevila, I., & Zarlenga, M. I. (2015). Smart city or smart citizens? The Barcelona case. *Journal of Strategy and Management, 8*(3). doi:10.1108/JSMA-03-2015-0030

Caprotti, F., & Liu, D. (2022). Platform urbanism and the Chinese smart city: The co-production and territorialisation of Hangzhou City Brain. *GeoJournal*, *87*(3). doi:10.1007/s10708-020-10320-2

Caragliu, A., & Del Bo, C. F. (2019). Smart innovative cities: The impact of Smart City policies on urban innovation. *Technological Forecasting and Social Change*, *142*(December 2017), 373–383. doi:10.1016/j.techfore.2018.07.022

Caragliu, A., & Del Bo, C. F. (2021). Smart cities and urban inequality. *Regional Studies*, (Mi), 1–32. doi:10.1080/00343404.2021.1984421

Caragliu, A., del Bo, C., & Nijkamp, P. (2011). Smart cities in Europe. *Journal of Urban Technology*, *18*(2), 65–82. doi:10.1080/10630732.2011.601117

Cardullo, P., & Kitchin, R. (2019). Smart urbanism and smart citizenship: The neoliberal logic of 'citizen-focused' smart cities in Europe. *Environment and Planning C: Politics and Space*, *37*(5). doi:10.1177/0263774X18806508

Carter, E., Adam, P., Tsakis, D., Shaw, S., Watson, R., & Ryan, P. (2020). Enhancing pedestrian mobility in Smart Cities using Big Data. *Journal of Management Analytics*, *7*(2). doi:10.1080/23270012.2020.1741039

Carvalho, L. (2015). Smart cities from scratch? A socio-technical perspective. *Cambridge Journal of Regions, Economy and Society*, *8*(1), 43–60. doi:10.1093/cjres/rsu010

Castelnovo, P., Morretta, V., & Vecchi, M. (2020). Regional disparities and industrial structure: Territorial capital and productivity in Italian firms. *Regional Studies*. doi:10.1080/00343404.2020.1763941

Chan, N. W. (2012). Economic and welfare impacts of disasters in East Asia and policy responses. *Economic and Welfare Impacts of Disasters in East Asia and Policy Responses*. doi:10.1007/978-4-431-55022-8

Chandler, A. D., & Saxenian, A. (1995). Regional advantage: Culture and competition in Silicon Valley and Route 128. *The New England Quarterly*, *68*(1). doi:10.2307/365972

Chang, I. C. C., Jou, S. C., & Chung, M. K. (2021). Provincialising smart urbanism in Taipei: The smart city as a strategy for urban regime transition. *Urban Studies*, *58*(3), 559–580. doi:10.1177/0042098020947908

Chatfield, A. T., & Reddick, C. G. (2019). A framework for Internet of Things-enabled smart government: A case of IoT cybersecurity policies and use cases in U.S. federal government. *Government Information Quarterly*, *36*(2), 346–357. doi:10.1016/j.giq.2018.09.007

Chatterjee, S., & Kar, A. K. (2015). Smart Cities in developing economies: A literature review and policy insights. In *2015 International Conference on Advances in Computing, Communications and Informatics, ICACCI*. doi:10.1109/ICACCI.2015.7275967

Chatterjee, S., Kar, A. K., & Gupta, M. P. (2018). Success of IoT in Smart Cities of India: An empirical analysis. *Government Information Quarterly*, *35*(3), 349–361. doi:10.1016/j.giq.2018.05.002

Chauhan, A., Jakhar, S. K., & Chauhan, C. (2021). The interplay of circular economy with industry 4.0 enabled smart city drivers of healthcare waste disposal. *Journal of Cleaner Production*, *279*, 123854. doi:10.1016/j.jclepro.2020.123854

Cheng, Y., Awan, U., Ahmad, S., & Tan, Z. (2021). How do technological innovation and fiscal decentralization affect the environment? A story of the fourth industrial revolution and sustainable growth. *Technological Forecasting and Social Change, 162*(August 2020), 120398. doi:10.1016/j.techfore.2020.120398

Chen, H., Gompers, P., Kovner, A., & Lerner, J. (2010). Buy local? The geography of venture capital. *Journal of Urban Economics*. doi:10.1016/j.jue.2009.09.013

Chia, S. Y. (2017). Singapore in 2016: Buffeted by external developments but preparing for the future. *Asian Survey, 57*. doi:10.1525/AS.2017.57.1.187

Christofi, M., Iaia, L., Marchesani, F., & Masciarelli, F. (2021). Marketing innovation and internationalization in smart city development: A systematic review, framework, and research agenda. *International Marketing Review, 38*(5), 948–984. doi:10.1108/IMR-01-2021-0027

Chu, Z., Cheng, M., & Yu, N. N. (2021). A smart city is a less polluted city. *Technological Forecasting and Social Change, 172*(April 2020), 121037. doi: 10.1016/j.techfore.2021.121037

Cimbaljević, M., Stankov, U., & Pavluković, V. (2019). Going beyond the traditional destination competitiveness–reflections on a smart destination in the current research. *Current Issues in Tourism, 22*(20), 2472–2477. doi:10.1080/13683500.2018.1529149

Clift, R., Sim, S., King, H., Chenoweth, J., Christie, I., Clavreul, J., … Murphy, R. (2017). The challenges of applying planetary boundaries as a basis for strategic decision-making in companies with global supply chains. *Sustainability, 9*(2), 279. doi:10.3390/su9020279

Coca-Stefaniak, J. A. (2021). Beyond smart tourism cities – Towards a new generation of “wise” tourism destinations. *Journal of Tourism Futures, 7*(2), 251–258. doi: 10.1108/JTF-11-2019-0130

Cocchia, A. (2014). Smart and digital city: A systematic literature review. doi:10.1007/978-3-319-06160-3_2

Cohen, B., Almirall, E., & Chesbrough, H. (2016). The city as a lab: Open innovation meets the collaborative economy. *California Management Review, 59*(1), 5–13. doi: 10.1177/0008125616683951

Colding, J., & Barthel, S. (2017). An urban ecology critique on the “Smart City” model. *Journal of Cleaner Production, 164*, 95–101. doi:10.1016/j.jclepro.2017.06.191

Collins, J., & Low, A. (2010). Asian female immigrant entrepreneurs in small and medium-sized businesses in Australia. *Entrepreneurship & Regional Development, 22*(1). doi:10.1080/08985620903220553

Contreras, G., & Platania, F. (2019). Economic and policy uncertainty in climate change mitigation: The London Smart City case scenario. *Technological Forecasting and Social Change, 142*. doi:10.1016/j.techfore.2018.07.018

Corburn, J., Asari, M. R., Pérez Jamarillo, J., & Gaviria, A. (2020). The transformation of Medellin into a ‘City for Life:’ insights for healthy cities. *Cities and Health, 4*(1). doi:10.1080/23748834.2019.1592735

Crevoisier, O., & Rime, D. (2021). Anchoring urban development: Globalisation, attractiveness and complexity. *Urban Studies, 58*(1), 36–52. doi:10.1177/0042098019889310

Criado, J. I., Sandoval-Almazan, R., & Gil-Garcia, J. R. (2013). Government innovation through social media. *Government Information Quarterly, 30*(4), 319–326. doi:10.1016/j.giq.2013.10.003

Curry, E. (2016). The big data value chain: Definitions, concepts, and theoretical approaches. In *New horizons for a data-driven economy: A roadmap for usage and exploitation of Big Data in Europe*. doi:10.1007/978-3-319-21569-3_3

Dameri, R. P., & Rosenthal-Sabroux, C. (2014). *Smart city and value creation* (pp. 1–12). doi:10.1007/978-3-319-06160-3_1

Datta, A. (2015). New urban utopias of postcolonial India: 'Entrepreneurial urbanization' in Dholera smart city, Gujarat. *Dialogues in Human Geography*, *5*(1), 3–22. doi:10.1177/2043820614565748

Datta, A., & Odendaal, N. (2019). Smart cities and the banality of power. *Environment and Planning D: Society and Space*, *37*(3), 387–392. doi:10.1177/0263775819841765

de Falco, S., Angelidou, M., & Addie, J. P. D. (2019). From the "smart city" to the "smart metropolis"? Building resilience in the urban periphery. *European Urban and Regional Studies*, *26*(2), 205–223. doi:10.1177/0969776418783813

De Guimarães, J. C. F., Severo, E. A., Felix Júnior, L. A., Da Costa, W. P. L. B., & Salmoria, F. T. (2020). Governance and quality of life in smart cities: Towards sustainable development goals. *Journal of Cleaner Production*, *253*. doi:10.1016/j.jclepro.2019.119926

De Jager, H. J., Mthembu, T. Z., Ngowi, A. B., & Chipunza, C. (2017). Towards an innovation and entrepreneurship ecosystem: A case study of the Central University of Technology, Free State. *Science, Technology and Society*. doi:10.1177/0971721817702292

De Silva, D. G., & McComb, R. P. (2012). Geographic concentration and high tech firm survival. *Regional Science and Urban Economics*. doi:10.1016/j.regsciurbeco.2012.03.001

De Simone, C., Ceci, F., & Alaimo, C. (2022). Data ecosystem and data value chain: An exploration of drones technology applications. In *Sustainable digital transformation: Paving the way towards smart organizations and societies* (pp. 203–218). Cham: Springer International Publishing.

Debarsy, N. (2012). The Mundlak approach in the spatial Durbin panel data model. *Spatial Economic Analysis*, *7*(1). doi:10.1080/17421772.2011.647059

Del Chiappa, G., Baggio, R., Wang, X., Li, X. R., Zhen, F., Zhang, J. H., … Hetman, O. (2016). Smart tourism destinations: Ecosystems for tourism destination competitiveness. *Tourism Management*, *4*(2), 145–150. doi:10.1016/j.tourman.2016.03.014

Dewi, M. A. A., Hidayanto, A. N., Purwandari, B., Kosandi, M., & Budi, N. F. A. (2018). Smart city readiness model based on technology-organization-environment (TOE) framework and its effect on adoption decision. In *Proceedings of the 22nd Pacific Asia Conference on Information Systems – Opportunities and Challenges for the Digitized Society: Are We Ready? PACIS 2018*, (Nathaniel 2014).

Dirks, S., & Keeling, M. (2009, June). *A vision of smarter cities: How cities can lead the way into a prosperous and sustainable future*. IBM Institute for Business Value.

Dobbelsteen, A. V. D., Broersma, S., Fremouw, M., Blom, T., Sturkenboom, J., & Martin, C. (2020). The Amsterdam energy transition roadmap – Introducing the City-zen methodology. *Smart and Sustainable Built Environment*, *9*(3). doi:10.1108/SASBE-05-2019-0065

Docherty, I., Marsden, G., & Anable, J. (2018). The governance of smart mobility. *Transportation Research Part A: Policy and Practice*. doi:10.1016/j.tra.2017.09.012

Dotti, N. F., Fratesi, U., Lenzi, C., & Percoco, M. (2014). Local labour market conditions and the spatial mobility of science and technology university students: Evidence from Italy. *Review of Regional Research*. doi:10.1007/s10037-014-0088-y

Drach-Zahavy, A. (2004). Exploring team support: The role of team's design, values, and leader's support. *Group Dynamics*. doi:10.1037/1089-2699.8.4.235

Echeverry Tamayo, J. D. (2019). Medellín: una ciudad de encrucijadas. Pobreza, modelo de ciudad y cambio social en el proyecto Cinturón Verde. *Territorios*, (40). doi:10.12804/revistas.urosario.edu.co/territorios/a.5535

El-Haddadeh, R., Weerakkody, V., Osmani, M., Thakker, D., & Kapoor, K. K. (2019). Examining citizens' perceived value of internet of things technologies in facilitating public sector services engagement. *Government Information Quarterly*, *36*(2), 310–320. doi:10.1016/j.giq.2018.09.009

Elhorst, J. P. (2003). Specification and estimation of spatial panel data models. *International Regional Science Review*, *26*(3). doi:10.1177/0160017603253791

Ellison, M., Bannister, J., Lee, W. D., & Haleem, M. S. (2021). Understanding policing demand and deployment through the lens of the city and with the application of big data. *Urban Studies*, *58*(15), 3157–3175. doi:10.1177/0042098020981007

English-Lueck, J. A., & Saveri, A. (2001). Silicon Missionaries and Identity Evangelists. *Anthropology of Work Review*, *22*(1). doi:10.1525/awr.2001.22.1.7

Evangelista, R., Guerrieri, P., & Meliciani, V. (2014). The economic impact of digital technologies in Europe. *Economics of Innovation and New Technology*, *23*(8), 802–824. doi:10.1080/10438599.2014.918438

Faggian, A., & McCann, P. (2009). Human capital and regional development. In *Handbook of regional growth and development theories*. doi:10.4337/9781788970020.00015

Feng, W., & Yuan, H. (2023). The impact of medical infrastructure on regional innovation: An empirical analysis of China's prefecture-level cities. *Technological Forecasting and Social Change*, *186*(PA), 122125. doi:10.1016/j.techfore.2022.122125

Fernandez-Anez, V., Fernández-Güell, J. M., & Giffinger, R. (2018). Smart City implementation and discourses: An integrated conceptual model. The case of Vienna. *Cities*, *78*(December 2017), 4–16. doi:10.1016/j.cities.2017.12.004

Florida, R. (1995). Toward the learning region. *Futures*, *27*(5), 527–536. doi:10.1016/0016-3287(95)00021-N

Florida, R. (2002a). The economic geography of talent. *Annals of the Association of American Geographers*, *92*(4), 743–755. doi:10.1111/1467-8306.00314

Florida, R. (2002b). The rise of the creative class. *Washington Monthly*, 15–26.

Florida, R. (2004). *Cities and the creative class*. doi:10.4324/9780203997673

Florida, R. (2012). *The Rise of the Creative Class-Revisited: 10th Anniversary Edition-Revised and Expanded*. Retrieved from http://www.amazon.com/The-Rise-Creative-Class-Revisited-Edition-Revised/dp/0465029930

Florida, R., Adler, P., & Mellander, C. (2017). The city as innovation machine. *Regional Studies*, *51*(1), 86–96. doi:10.1080/00343404.2016.1255324

Florida, R., & Kenney, M. (1992). Restructuring in place: Japanese investment, production organization, and the geography of steel. *Economic Geography*. doi:10.2307/144199

Fortino, G., Russo, W., Savaglio, C., Shen, W., & Zhou, M. (2018). Agent-oriented cooperative smart objects: From IoT system design to implementation. *IEEE Transactions on Systems, Man, and Cybernetics: Systems*, *48*(11), 1949–1956. doi: 10.1109/TSMC.2017.2780618

Freeman, R. B. (2010). Globalization of scientific and engineering talent: International mobility of students, workers, and ideas and the world economy. *Economics of Innovation and New Technology*. doi:10.1080/10438590903432871

Fujita, M., & Thisse, J. F. (2003). Agglomeration and market interaction. In *Advances in economics and econometrics: Theory and applications, Eighth World Congress* (Vol. 1). doi:10.1017/CBO9780511610240.011

Gafter, L. M., & Tchetchik, A. (2017). The role of social ties and communication technologies in visiting friends tourism-A GMM simultaneous equations approach. *Tourism Management*, *61*, 343–353. doi:10.1016/j.tourman.2017.02.024

Gagliardi, D., Schina, L., Sarcinella, M. L., Mangialardi, G., Niglia, F., & Corallo, A. (2017). Information and communication technologies and public participation: Interactive maps and value added for citizens. *Government Information Quarterly*, *34*(1), 153–166. doi:10.1016/j.giq.2016.09.002

Gaur, A., Scotney, B., Parr, G., & McClean, S. (2015). Smart city architecture and its applications based on IoT. *Procedia Computer Science*, *52*(1). doi:10.1016/j.procs.2015.05.122

Giarda, E., & Moroni, G. (2018). The degree of poverty persistence and the role of regional disparities in Italy in comparison with France, Spain and the UK. *Social Indicators Research*, *136*(1), 163–202. doi:10.1007/s11205-016-1547-3

Giffinger, R., Fertner, C., Kramar, H., & Meijers, E. (2007). *City-ranking of European medium-sized cities*. Vienna, UT: Centre of Regional Science.

Gil-Garcia, J. R., Dawes, S. S., & Pardo, T. A. (2018). Digital government and public management research: Finding the crossroads. *Public Management Review*, *20*(5), 633–646. doi:10.1080/14719037.2017.1327181

Gil-Garcia, J. R., Zhang, J., & Puron-Cid, G. (2016). Conceptualizing smartness in government: An integrative and multi-dimensional view. *Government Information Quarterly*, *33*(3), 524–534. doi:10.1016/j.giq.2016.03.002

Glaeser, E. L. (1999). Learning in cities. *Journal of Urban Economics*, *46*(2). doi: 10.1006/juec.1998.2121

Glaeser, E. L. (2000). The new economics of urban and regional growth. *The American Economic Review*.

Glaeser, E. L., & Gottlieb, J. D. (2009). The wealth of cities: Agglomeration economies and spatial equilibrium in the United States. *Journal of Economic Literature*, *47*(4). doi:10.1257/jel.47.4.983

Global Data Thematic Research. (2020). History of smart cities: Timeline. *Verdict.Co.Uk*.

Goldstein, H., & Drucker, J. (2006). The economic development impacts of universities on regions: Do size and distance matter? *Economic Development Quarterly*. doi:10.1177/0891242405283387

Goodspeed, R. (2015). Smart cities: Moving beyond urban cybernetics to tackle wicked problems. *Cambridge Journal of Regions, Economy and Society*, *8*(1), 79–92. doi:10.1093/cjres/rsu013

Greenwood, M. J., & Sweetland, D. (1972). The determinants of migration between standard metropolitan statistical areas. *Demography*. doi:10.2307/2060673

Gretzel, U., & Koo, C. (2021). Smart tourism cities: A duality of place where technology supports the convergence of touristic and residential experiences. *Asia Pacific Journal of Tourism Research*. doi:10.1080/10941665.2021.1897636

Gretzel, U., Sigala, M., Xiang, Z., & Koo, C. (2015). Smart tourism: Foundations and developments. *Electronic Markets*, *25*(3). doi:10.1007/s12525-015-0196-8

Grimaldi, D., & Fernandez, V. (2017). The alignment of University curricula with the building of a Smart City: A case study from Barcelona. *Technological Forecasting and Social Change*, *123*, 298–306. doi:10.1016/j.techfore.2016.03.011

Grossi, G., & Pianezzi, D. (2017). Smart cities: Utopia or neoliberal ideology? *Cities*, *69*(July), 79–85. doi:10.1016/j.cities.2017.07.012

Große-Bley, J., & Kostka, G. (2021). Big Data dreams and reality in Shenzhen: An investigation of Smart City implementation in China. *Big Data and Society*, *8*(2). doi:10.1177/20539517211045171

Guzman, J., & Kacperczyk, A. (Olenka). (2019). Gender gap in entrepreneurship. *Research Policy*, *48*(7), 1666–1680. doi:10.1016/j.respol.2019.03.012

Hajer, M., & Bröer, C. (2020). We are here! Claim-making and Claim-placing of Undocumented Migrants in Amsterdam. *European Journal of Cultural and Political Sociology*, *7*(4). doi:10.1080/23254823.2020.1774911

Hall, T., & Hubbard, P. (1998). *The entrepreneurial city: Geographies of politics, regime and representation*. Chichester: Wiley.

Hämäläinen, M. (2020). A framework for a smart city design: Digital transformation in the Helsinki Smart City. In *Contributions to management science*. doi:10.1007/978-3-030-23604-5_5

Harrison, C., Eckman, B., Hamilton, R., Hartswick, P., Kalagnanam, J., Paraszczak, J., & Williams, P. (2010). Foundations for smarter cities. *IBM Journal of Research and Development*, *54*(4). doi:10.1147/JRD.2010.2048257

Harvey, D. (1989). *The condition of postmodernity: An enquiry into the origins of cultural change*. doi:10.2307/2072256

Hashem, I. A. T., Chang, V., Anuar, N. B., Adewole, K., Yaqoob, I., Gani, A., … Chiroma, H. (2016). The role of big data in smart city. *International Journal of Information Management*, *36*(5), 748–758. doi:10.1016/j.ijinfomgt.2016.05.002

Hofstede, G. (2011). Dimensionalizing cultures: The Hofstede model in context. *Online Readings in Psychology and Culture*. doi:10.9707/2307-0919.1014

Holbeche, L. S. (2018). Organisational effectiveness and agility. *Journal of Organizational Effectiveness*, *5*(4). doi:10.1108/JOEPP-07-2018-0044

Hollands, R. G. (2008). Will the real smart city please stand up? Intelligent, progressive or entrepreneurial? *City*, *12*(3), 303–320. doi:10.1080/13604810802479126

Hollands, R. G. (2015). Critical interventions into the corporate smart city. *Cambridge Journal of Regions, Economy and Society*, *8*(1), 61–77. doi:10.1093/cjres/rsu011

Hollands, R. G. (2020). Will the real smart city please stand up? In *The Routledge companion to smart cities*. doi:10.4324/9781315178387-13

Holmes, T. J., Fujita, M., Krugman, P., & Venables, A. J. (2000). The spatial economy: Cities, regions, and international trade. *Southern Economic Journal*. doi:10.2307/1061487

Huertas, A., Moreno, A., & Pascual, J. (2021). Place branding for smart cities and smart tourism destinations: Do they communicate their smartness? *Sustainability (Switzerland)*, *13*(19), 1–18. doi:10.3390/su131910953

Ibrahim, M., El-Zaart, A., & Adams, C. (2018). Smart sustainable cities roadmap: Readiness for transformation towards urban sustainability. *Sustainable Cities and Society*, *37*(March 2017), 530–540. doi:10.1016/j.scs.2017.10.008

Irazábal, C., & Jirón, P. (2021). Latin American smart cities: Between worlding infatuation and crawling provincialising. *Urban Studies*, *58*(3), 507–534. doi: 10.1177/0042098020945201

Jacobs, J. (1961). *The death and life of great American cities*. New York, NY: The Failure of Town Planning. doi:10.2307/794509

Jacobs, J. (1969). Cities first – Rural development later. *The Economy of Cities*. 3–48

Jindal, A., Kumar, N., & Singh, M. (2020). A unified framework for big data acquisition, storage, and analytics for demand response management in smart cities. *Future Generation Computer Systems*, *108*, 921–934. doi:10.1016/j.future.2018.02.039

Jirón, P., Imilán, W. A., Lange, C., & Mansilla, P. (2021). Placebo urban interventions: Observing Smart City narratives in Santiago de Chile. *Urban Studies*, *58*(3), 601–620. doi:10.1177/0042098020943426

Jovicic, D. Z. (2019). From the traditional understanding of tourism destination to the smart tourism destination. *Current Issues in Tourism*, *22*(3), 276–282. doi:10.1080/13683500.2017.1313203

Kapoor, R. (2018). Ecosystems: Broadening the locus of value creation. *Journal of Organ Dysfunction*, *7*(1). doi:10.1186/s41469-018-0035-4

Kar, A. K., Ilavarasan, V., Gupta, M. P., Janssen, M., & Kothari, R. (2019). Moving beyond smart cities: Digital nations for social innovation & sustainability. *Information Systems Frontiers*, *21*(3), 495–501. doi:10.1007/s10796-019-09930-0

Karimikia, H., Bradshaw, R., Singh, H., Ojo, A., Donnellan, B., & Guerin, M. (2022). An emergent taxonomy of boundary spanning in the smart city context – The case of smart Dublin. *Technological Forecasting and Social Change*, *185*(February), 122100. doi:10.1016/j.techfore.2022.122100

Khan, M. S., Woo, M., Nam, K., & Chathoth, P. K. (2017). Smart city and smart tourism: A case of Dubai. *Sustainability (Switzerland)*, *9*(12). doi:10.3390/su9122279

Khatoun, R., & Zeadally, S. (2016). Smart cities: Concepts, architectures, research opportunities. *Communications of the ACM*, *59*(8). doi:10.1145/2858789

Kimbu, A. N., & Ngoasong, M. Z. (2016). Women as vectors of social entrepreneurship. *Annals of Tourism Research*, *60*. doi:10.1016/j.annals.2016.06.002

Kiss, M., & Muha, L. (2018). The cybersecurity capability aspects of smart government and Industry 4.0 programmes. *Interdisciplinary Description of Complex Systems*, *16*(3), 313–319. doi:10.7906/indecs.16.3.2

Knudsen, B., Florida, R., Gates, G., & Stolarick, K. (2007, May). Urban density, creativity and innovation.

Kogut, C. S., & Mejri, K. (2022). Female entrepreneurship in emerging markets: Challenges of running a business in turbulent contexts and times. *International Journal of Gender and Entrepreneurship*, *14*(1), 95–116. doi:10.1108/IJGE-03-2021-0052

Krabokoukis, T., & Polyzos, S. (2020). An investigation of factors determining the tourism attractiveness of Greece's prefectures. *Journal of the Knowledge Economy*. doi:10.1007/s13132-020-00704-7

Krishnan, B., Arumugam, S., & Maddulety, K. (2020). Critical success factors for the digitalization of smart cities. *International Journal of Technology Management and Sustainable Development*, *19*(1), 69–86. doi:10.1386/tmsd_00016_1

Kumar, M., Graham, G., Hennelly, P., & Srai, J. (2016). How will smart city production systems transform supply chain design: A product-level investigation. *International Journal of Production Research*, *54*(23), 7181–7192. doi:10.1080/00207543.2016.1198057

Kumar, S., Mookerjee, V., & Shubham, A. (2018). Research in operations management and information systems interface. *Production and Operations Management*, *27*(11), 1893–1905. doi:10.1111/poms.12961

Kummitha, R. K. R. (2018). Entrepreneurial urbanism and technological panacea: Why Smart City planning needs to go beyond corporate visioning? *Technological Forecasting and Social Change*, *137*(September 2017), 330–339. doi:10.1016/j.techfore.2018.07.010

Kummitha, R. K. R. (2019). Smart cities and entrepreneurship: An agenda for future research. *Technological Forecasting and Social Change*, *149*(July 2018), 119763. doi:10.1016/j.techfore.2019.119763

Kummitha, R. K. R. (2020). Why distance matters: The relatedness between technology development and its appropriation in smart cities. *Technological Forecasting and Social Change*, *157*(April), 120087. doi:10.1016/j.techfore.2020.120087

Kummitha, R. K. R., & Crutzen, N. (2017). How do we understand smart cities? An evolutionary perspective. *Cities*, *67*(March), 43–52. doi:10.1016/j.cities.2017.04.010

Kummitha, R. K. R., & Crutzen, N. (2019). Smart cities and the citizen-driven internet of things: A qualitative inquiry into an emerging smart city. *Technological Forecasting and Social Change*, *140*(October 2018), 44–53. doi:10.1016/j.techfore.2018.12.001

Langley, P., & Leyshon, A. (2017). Platform capitalism: The intermediation and capitalization of digital economic circulation. *Finance and Society*. doi:10.2218/finsoc.v3i1.1936

Laursen, K., Masciarelli, F., & Prencipe, A. (2012). Trapped or spurred by the home region the effects of potential social capital on involvement in foreign markets for goods and technology. *Journal of International Business Studies*, *43*(9), 783–807. doi:10.1057/jibs.2012.27

Laursen, K., Masciarelli, F., & Reichstein, T. (2016). A matter of location: The role of regional social capital in overcoming the liability of newness in R&D acquisition activities. *Regional Studies*, *50*(9), 1537–1550. doi:10.1080/00343404.2015.1041370

Lee, S. Y., Florida, R., & Acs, Z. J. (2004). Creativity and entrepreneurship: A regional analysis of new firm formation. *Regional Studies*. doi:10.1080/0034340042000280910.

Lee, S. Y., Florida, R., & Gates, G. (2010). Innovation, human capital, and creativity. *International Review of Psycho-Analysis*. doi:10.1080/12294659.2010.10805158

Lee, J. H., Hancock, M. G., & Hu, M. C. (2014). Towards an effective framework for building smart cities: Lessons from Seoul and San Francisco. *Technological Forecasting and Social Change*, *89*. doi:10.1016/j.techfore.2013.08.033

Lee, J., & Lee, H. (2014). Developing and validating a citizen-centric typology for smart city services. *Government Information Quarterly*, *31*(Suppl. 1), S93–S105. doi:10.1016/j.giq.2014.01.010

Lee, J., Yang, H., Lee, B., Cho, A., & Oh, I. (2016). Design research direction for smart mobility in mega cities. *Archives of Design Research, 29*(3). doi:10.15187/adr.2016.08.29.3.63

Lee, L. F., & Yu, J. (2014). Efficient GMM estimation of spatial dynamic panel data models with fixed effects. *Journal of Econometrics, 180*(2). doi:10.1016/j.jeconom.2014.03.003

Leitheiser, S., & Follmann, A. (2020). The social innovation–(re)politicisation nexus: Unlocking the political in actually existing smart city campaigns? The case of SmartCity Cologne, Germany. *Urban Studies, 57*(4), 894–915. doi:10.1177/0042098019869820

Lei, S. I., Ye, S., Wang, D., & Law, R. (2020). Engaging customers in value co-creation through mobile instant messaging in the tourism and hospitality industry. *Journal of Hospitality & Tourism Research, 44*(2), 229–251. doi:10.1177/1096348019893066

Leonelli, S., Marchesani, F., & Masciarelli, F. (2022). Risk or opportunity? Exploring the relationship between entrepreneurial decision and the use of equity crowdfunding campaigns in less-and well-developed regions in Ital. In *The international dimension of entrepreneurial decision-making* (pp. 99–114). doi:10.1016/S0140-6736(46)91888-0

Leung, D., Law, R., van Hoof, H., & Buhalis, D. (2013). Social media in tourism and hospitality: A literature review. *Journal of Travel and Tourism Marketing, 30*(1–2), 3–22. doi:10.1080/10548408.2013.750919

Levy, P. S., & Lemeshow, S. (2011). *Sampling of populations: Methods and applications* (4th ed.). doi:10.1002/9780470374597

Li, Z., & Liao, Q. (2018). Economic solutions to improve cybersecurity of governments and smart cities via vulnerability markets. *Government Information Quarterly, 35*(1), 151–160. doi:10.1016/j.giq.2017.10.006

Lim, Y., Edelenbos, J., & Gianoli, A. (2019). Identifying the results of smart city development: Findings from systematic literature review. *Cities, 95*(June), 102397. doi:10.1016/j.cities.2019.102397

Linders, D. (2012). From e-government to we-government: Defining a typology for citizen coproduction in the age of social media. *Government Information Quarterly, 29*(4), 446–454. doi:10.1016/j.giq.2012.06.003

Linde, L., Sjödin, D., Parida, V., & Wincent, J. (2021). Dynamic capabilities for ecosystem orchestration: A capability-based framework for smart city innovation initiatives. *Technological Forecasting and Social Change, 166*. doi:10.1016/j.techfore.2021.120614

Liu, X., Yang, M., & Nie, X. (2023). Can city brand reduce urban air pollution?—An empirical research based on "National Civilized City" in China. *Technological Forecasting and Social Change, 186*(PB), 122179. doi:10.1016/j.techfore.2022.122179

Li, G., Wang, Y., Luo, J., & Li, Y. (2018). Evaluation on construction level of smart city: An empirical study from Twenty Chinese cities. *Sustainability (Switzerland), 10*(9). doi:10.3390/su10093348

Lnenicka, M., Nikiforova, A., Luterek, M., Azeroual, O., Ukpabi, D., Valtenbergs, V., & Machova, R. (2022). Transparency of open data ecosystems in smart cities: Definition and assessment of the maturity of transparency in 22 smart cities. *Sustainable Cities and Society, 82*(July). doi:10.1016/j.scs.2022.103906

Löfgren, K., & Webster, C. W. R. (2020). The value of Big Data in government: The case of 'smart cities'. *Big Data and Society*, *7*(1). doi:10.1177/2053951720912775

Lytras, M. D., & Visvizi, A. (2018). Who uses smart city services and what to make of it: Toward interdisciplinary smart cities research. *Sustainability (Switzerland)*. doi: 10.3390/su10061998

Magazzini, L., & Calzolari, G. (2020). Testing initial conditions in dynamic panel data models. *Econometric Reviews*, *39*(2). doi:10.1080/07474938.2019.1690194

Mangiaracina, R., Perego, A., Salvadori, G., & Tumino, A. (2017). A comprehensive view of intelligent transport systems for urban smart mobility. *International Journal of Logistics Research and Applications*, *20*(1), 39–52. doi:10.1080/13675567.2016.1241220

Marchesani, F. (2022). Exploring the relationship between digital services advance and smart tourism in cities: Empirical evidence from Italy. *Current Issues in Tourism*, 1–12. doi:10.1080/13683500.2022.2153652

Marchesani, F., Iaia, L., Masciarelli, F., & Christofi, M. (2022). Smart City's Internationalization and International Management Strategies in the Digital Era: A Systematic Literature Review. *Sustainable Digital Transformation*, 153–165.

Marchesani, F., & Masciarelli, F. (2021). *Crowdfunding as entrepreneurial investment: The role of local knowledge spillover* (pp. 92–108). doi:10.1007/978-3-030-87842-9_8

Marchesani, F., & Masciarelli, F. (2023). Does tourism flow in cities drive green practices in the current smart city trajectories? Empirical evidence from Italy. *International Journal of Tourism Cities*, *9*(3).

Marchesani, F., Masciarelli, F., & Bikfalvi, A. (2023). Smart city as a hub for talent and innovative companies: Exploring the (dis) advantages of digital technology implementation in cities. *Technological Forecasting and Social Change*, *193*(May), 122636. doi:10.1016/j.techfore.2023.122636

Marchesani, F., Masciarelli, F., & Doan, H. Q. (2022). Innovation in cities a driving force for knowledge flows: Exploring the relationship between high-tech firms, student mobility, and the role of youth entrepreneurship. *Cities*, *130*, 103852. doi: 10.1016/J.CITIES.2022.103852

March, H., & Ribera-Fumaz, R. (2016). Smart contradictions: The politics of making Barcelona a Self-sufficient city. *European Urban and Regional Studies*, *23*(4), 816–830. doi:10.1177/0969776414554488

Mariani, M. M., Di Felice, M., & Mura, M. (2016). Facebook as a destination marketing tool: Evidence from Italian regional Destination Management Organizations. *Tourism Management*, *54*, 321–343. doi:10.1016/j.tourman.2015.12.008

Marsal-Llacuna, M. L., Colomer-Llinàs, J., & Meléndez-Frigola, J. (2015). Lessons in urban monitoring taken from sustainable and livable cities to better address the Smart Cities initiative. *Technological Forecasting and Social Change*, *90*(PB). doi: 10.1016/j.techfore.2014.01.012

Marsal-Llacuna, M. L., & Segal, M. E. (2017). The Intelligenter Method (II) for "smarter" urban policy-making and regulation drafting. *Cities*, *61*. doi:10.1016/j.cities.2016.05.006

Martin, C. J., Evans, J., & Karvonen, A. (2018). Smart and sustainable? Five tensions in the visions and practices of the smart-sustainable city in Europe and North America. *Technological Forecasting and Social Change*, *133*(February), 269–278. doi:10.1016/j.techfore.2018.01.005

Martínez-Rodríguez, I., Quintana-Rojo, C., Gento, P., & Callejas-Albiñana, F. E. (2022). Public policy recommendations for promoting female entrepreneurship in Europe. *The International Entrepreneurship and Management Journal*, *18*(3), 1235–1262. doi:10.1007/s11365-021-00751-9

Masik, G., Sagan, I., & Scott, J. W. (2021). Smart City strategies and new urban development policies in the Polish context. *Cities*, *108*(June 2020), 102970. doi: 10.1016/j.cities.2020.102970

Massana, J., Pous, C., Burgas, L., Melendez, J., & Colomer, J. (2017). Identifying services for short-term load forecasting using data driven models in a Smart City platform. *Sustainable Cities and Society*, *28*. doi:10.1016/j.scs.2016.09.001

Matos, F., Vairinhos, V. M., Dameri, R. P., & Durst, S. (2017). Increasing smart city competitiveness and sustainability through managing structural capital. *Journal of Intellectual Capital*, *18*(3), 693–707. doi:10.1108/JIC-12-2016-0141

Meijer, A., & Bolívar, M. P. R. (2016). Governing the smart city: A review of the literature on smart urban governance. *International Review of Administrative Sciences*. doi:10.1177/0020852314564308

Mergel, I., Edelmann, N., & Haug, N. (2019). Defining digital transformation: Results from expert interviews. *Government Information Quarterly*, *36*(4). doi:10.1016/j.giq.2019.06.002

Morgan, K., & Webb, B. (2020). Googling the city: In search of the public interest on Toronto's 'Smart' waterfront. *Urban Planning*, *5*(1), 84–95.

Morinière, L. (2012). Environmentally influenced urbanisation: Footprints bound for town? *Urban Studies*, *49*(2), 435–450. doi:10.1177/0042098011402233

Mossberger, K., Wu, Y., & Crawford, J. (2013). Connecting citizens and local governments? Social media and interactivity in major U.S. cities. *Government Information Quarterly*, *30*(4), 351–358. doi:10.1016/j.giq.2013.05.016

Mouton, M., & Burns, R. (2021). (Digital) neo-colonialism in the smart city. *Regional Studies*, *55*(12), 1890–1901. doi:10.1080/00343404.2021.1915974

Muñoz, P., & Cohen, B. (2016). The making of the urban entrepreneur. *California Management Review*, *59*(1), 71–91. doi:10.1177/0008125616683953

Nakano, S., & Washizu, A. (2021). Will smart cities enhance the social capital of residents? The importance of smart neighborhood management. *Cities*, *115*, 103244. doi:10.1016/j.cities.2021.103244

Nam, T., & Pardo, T. A. (2011a). Conceptualizing smart city with dimensions of technology, people, and institutions. In *ACM International Conference Proceeding Series*. doi:10.1145/2037556.2037602

Nam, T., & Pardo, T. A. (2011b). Smart city as urban innovation: Focusing on management, policy, and context. In *ACM International Conference Proceeding Series*. doi:10.1145/2072069.2072100

Nam, T., & Pardo, T. A. (2014). The changing face of a city government: A case study of Philly311. *Government Information Quarterly*, *31*(suppl. 1), S1–S9. doi:10.1016/j.giq.2014.01.002

Nanda, R., & Rhodes-Kropf, M. (2013). Investment cycles and startup innovation. *Journal of Financial Economics*. doi:10.1016/j.jfineco.2013.07.001

Nasir, M. A., Duc Huynh, T. L., & Xuan Tram, H. T. (2019). Role of financial development, economic growth & foreign direct investment in driving climate change: A case of emerging ASEAN. *Journal of Environmental Management*, *242*(January), 131–141. doi:10.1016/j.jenvman.2019.03.112

Naudé, W., Gries, T., Wood, E., & Meintjies, A. (2008). Regional determinants of entrepreneurial start-ups in a developing country. *Entrepreneurship and Regional Development*. doi:10.1080/08985620701631498

Neirotti, P., De Marco, A., Cagliano, A. C., Mangano, G., & Scorrano, F. (2014). Current trends in smart city initiatives: Some stylised facts. *Cities*. doi:10.1016/j.cities.2013.12.010

Nesti, G. (2019). Mainstreaming gender equality in smart cities: Theoretical, methodological and empirical challenges. *Information Polity*, *24*(3). doi:10.3233/IP-190134

Nesti, G., & Graziano, P. R. (2020). The democratic anchorage of governance networks in smart cities: An empirical assessment. *Public Management Review*, *22*(5), 648–667. doi:10.1080/14719037.2019.1588355

Neumann, O., Matt, C., Hitz-Gamper, B. S., Schmidthuber, L., & Stürmer, M. (2019). Joining forces for public value creation? Exploring collaborative innovation in smart city initiatives. *Government Information Quarterly*, *36*(4), 101411. doi:10.1016/j.giq.2019.101411

Nevola, F., Coles, T., & Mosconi, C. (2022). Hidden Florence revealed? Critical insights from the operation of an augmented reality app in a World Heritage City. *Journal of Heritage Tourism*, *17*(4). doi:10.1080/1743873X.2022.2036165

Nguyen, B. (2021). Regional informal institutions, local governance and gendered entrepreneurship. *Regional Studies*, *55*(7), 1169–1181. doi:10.1080/00343404.2021.1889489

Ng, I. C. L., & Wakenshaw, S. Y. L. (2017). The Internet-of-Things: Review and research directions. *International Journal of Research in Marketing*, *34*(1), 3–21. doi:10.1016/j.ijresmar.2016.11.003

Nieto, M., & Quevedo, P. (2005). Absorptive capacity, technological opportunity, knowledge spillovers, and innovative effort. *Technovation*. doi:10.1016/j.technovation.2004.05.001

Nikki Han, M. J., & Kim, M. J. (2021). A critical review of the smart city in relation to citizen adoption towards sustainable smart living. *Habitat International*, *108*(January), 102312. doi:10.1016/j.habitatint.2021.102312

Nilsson, J. H. (2019). Urban bicycle tourism: Path dependencies and innovation in Greater Copenhagen. *Journal of Sustainable Tourism*, *27*(11), 1648–1662. doi:10.1080/09669582.2019.1650749

Noori, N., de Jong, M., & Hoppe, T. (2020). Towards an integrated framework to measure smart city readiness: The case of Iranian cities. *Smart Cities*, *3*(3), 676–704. doi:10.3390/smartcities3030035

Noori, N., de Jong, M., Janssen, M., Schraven, D., & Hoppe, T. (2021). Input-output modeling for smart city development. *Journal of Urban Technology*, *28*(1–2), 71–92. doi:10.1080/10630732.2020.1794728

Odendaal, N. (2021). Everyday urbanisms and the importance of place: Exploring the elements of the emancipatory smart city. *Urban Studies*, *58*(3), 639–654. doi:10.1177/0042098020970970

Ooms, W., Caniëls, M. C. J., Roijakkers, N., & Cobben, D. (2020). Ecosystems for smart cities: Tracing the evolution of governance structures in a Dutch smart city initiative. *The International Entrepreneurship and Management Journal*, *16*(4). doi:10.1007/s11365-020-00640-7

Orlowski, A., & Romanowska, P. (2019). Smart Cities Concept: Smart Mobility Indicator. *Cybernetics & Systems, 50*(2), 118–131. doi:10.1080/01969722.2019.1565120

Oztemel, E., & Gursev, S. (2020). Literature review of Industry 4.0 and related technologies. *Journal of Intelligent Manufacturing, 31*(1), 127–182. doi:10.1007/s10845-018-1433-8

Pan, J. G., Lin, Y. F., Chuang, S. Y., & Kao, Y. C. (2011). From governance to service-smart city evaluations in Taiwan. In *Proceedings – 2011 International Joint Conference on Service Sciences, IJCSS 2011* (pp. 334–337). doi:10.1109/IJCSS.2011.74

Paskaleva, K. A. (2009). Enabling the smart city: The progress of city e-governance in Europe. *International Journal of Innovation and Regional Development.* doi:10.1504/ijird.2009.022730

Paskaleva, K. A. (2011). The smart city: A nexus for open innovation? *Intelligent Buildings International.* doi:10.1080/17508975.2011.586672

Paskaleva, K., & Cooper, I. (2018). Open innovation and the evaluation of internet-enabled public services in smart cities. *Technovation, 78*(July), 4–14. doi:10.1016/j.technovation.2018.07.003

Pereira, G. V., Parycek, P., Falco, E., & Kleinhans, R. (2018). Smart governance in the context of smart cities: A literature review. *Information Polity, 23*(2), 143–162. doi:10.3233/IP-170067

Perry, T. S. (2018). San Diego's streetlights get smart. *IEEE Spectrum, 55.* doi:10.1109/MSPEC.2018.8241729

Pinna, F., Masala, F., & Garau, C. (2017). Urban policies and mobility trends in Italian smart cities. *Sustainability (Switzerland), 9*(4). doi:10.3390/su9040494

Pittaway, J. J., & Montazemi, A. R. (2020). Know-how to lead digital transformation: The case of local governments. *Government Information Quarterly.* doi:10.1016/j.giq.2020.101474

Poggesi, S., Mari, M., & De Vita, L. (2016). What's new in female entrepreneurship research? Answers from the literature. *The International Entrepreneurship and Management Journal, 12*(3), 735–764. doi:10.1007/s11365-015-0364-5

Porter, M. E. (1996). Competitive advantage, agglomeration economies, and regional policy. *International Regional Science Review, 19*(2). doi:10.1177/016001769601900208

Porter, M. (1998). The competitive advantage of nations: With a new introduction. *Harvard Business Review.* doi:10.1016/j.technovation.2007.06.002

Poulhès, A., & Proulhac, L. (2021). The Paris Region low emission zone, a benefit shared with residents outside the zone. *Transportation Research Part D: Transport and Environment, 98.* doi:10.1016/j.trd.2021.102977

Putnam, R. D. (2000). *Bowling alone: The collapse and revival of American Community*, 2001. ISBN. Policy Analysis, 20. New York, NY: Simon and Schuster.

Qi, W., & Shen, Z. J. M. (2019). A Smart-City scope of operations management. *Production and Operations Management, 28*(2), 393–406. doi:10.1111/poms.12928

Qonita, M., & Giyarsih, S. R. (2022). Smart city assessment using the Boyd Cohen smart city wheel in Salatiga, Indonesia. *GeoJournal.* doi:10.1007/s10708-022-10614-7

Rahman, M. A., Rashid, M. M., Shamim Hossain, M., Hassanain, E., Alhamid, M. F., & Guizani, M. (2019). Blockchain and IoT-based cognitive edge framework for sharing economy services in a smart city. *IEEE Access*, *7*, 18611–18621. doi: 10.1109/ACCESS.2019.2896065

Reddick, C. G., Enriquez, R., Harris, R. J., & Sharma, B. (2020). Determinants of broadband access and affordability: An analysis of a community survey on the digital divide. *Cities*, *106*(September), 102904. doi:10.1016/j.cities.2020.102904

Reinartz, W., Wiegand, N., & Imschloss, M. (2019). The impact of digital transformation on the retailing value chain. *International Journal of Research in Marketing*. doi:10.1016/j.ijresmar.2018.12.002

Rohracher, H., & Späth, P. (2014). The interplay of urban energy policy and socio-technical transitions: The eco-cities of Graz and Freiburg in retrospect. *Urban Studies*, *51*(7). doi:10.1177/0042098013500360

Romão, J., Kourtit, K., Neuts, B., & Nijkamp, P. (2018). The smart city as a common place for tourists and residents: A structural analysis of the determinants of urban attractiveness. *Cities*, *78*(November 2017), 67–75. doi:10.1016/j.cities.2017.11.007

Roodman, D. (2009a). A note on the theme of too many instruments. *Oxford Bulletin of Economics & Statistics*, *71*(1). doi:10.1111/j.1468-0084.2008.00542.x

Roodman, D. (2009b). How to do xtabond2: An introduction to difference and system GMM in Stata. *Stata Journal*, *9*(1). doi:10.1177/1536867x0900900106

Roodman, D. (2009c). Practitioners' corner: A note on the theme of too many instruments. *Oxford Bulletin of Economics & Statistics*, *71*(1). doi:10.1111/j.1468-0084.2008.00542.x

Rosca, E., Agarwal, N., & Brem, A. (2020). Women entrepreneurs as agents of change: A comparative analysis of social entrepreneurship processes in emerging markets. *Technological Forecasting and Social Change*, *157*(November 2018), 120067. doi:10.1016/j.techfore.2020.120067

Rosenthal, S. S., & Strange, W. C. (2003). Geography, industrial organization, and agglomeration. *The Review of Economics and Statistics*. doi:10.1162/003465303765299882

Salom-Carrasco, J., Pitarch-Garrido, M. D., & Sales-Ten, A. (2017). Social innovation: Urban strategies in a context of change. The case of the city of Valencia. *CIRIEC – España. Revista de Economía Pública, Social y Cooperativa*, *0*(91). doi:10.7203/CIRIEC-E.91.10451

Salvia, G., & Morello, E. (2020). Sharing cities and citizens sharing: Perceptions and practices in Milan. *Cities*, *98*. doi:10.1016/j.cities.2019.102592

Sancino, A., & Hudson, L. (2020). Leadership in, of, and for smart cities–case studies from Europe, America, and Australia. *Public Management Review*, *22*(5), 701–725. doi:10.1080/14719037.2020.1718189

Sargan, J. D. (1958). The estimation of economic relationships using instrumental variables. *Econometrica*, *26*(3). doi:10.2307/1907619

Sassen, S. (2013). The global city: New York, London, Tokyo. doi:10.2307/2152688

Saxenian, A. L. (2002). Silicon Valley's new immigrant high-growth entrepreneurs. *Economic Development Quarterly*, *16*(1). doi:10.1177/0891242402016001003

Schaefer, M., & Hetman, O. (2019). Effective tools of digital marketing implementation. *University Economic Bulletin*, *41*, 67–74. doi:10.31470/2306-546x-2019-41-67-74

Scholl, H. J., & Alawadhi, S. (2016). Creating Smart Governance: The key to radical ICT overhaul at the City of Munich. *Information Polity*, *21*(1). doi:10.3233/IP-150369

Schoof, U. (International L. O.) (2006). *Stimulating Youth Entrepreneurship: Barriers and incentives to enterprise start-ups by young people*. Youth and Entrepreneurship.

Schultz, T. W. (1961). Investment in human capital: Reply. *The American Economic Review*. doi:10.2307/1813848

Scornavacca, E., Paolone, F., Za, S., & Martiniello, L. (2020). Investigating the entrepreneurial perspective in smart city studies. *The International Entrepreneurship and Management Journal*, *16*(4), 1197–1223. doi:10.1007/s11365-019-00630-4

Setijadi, E., Darmawan, A. K., Mardiyanto, R., Santosa, I., Hoiriyah, & Kristanto, T. (2019). A model for evaluation smart city readiness using structural equation modelling: A citizen's perspective. In *Proceedings of 2019 4th International Conference on Informatics and Computing, ICIC 2019*. doi:10.1109/ICIC47613.2019.8985969

Seyfi, S., & Hall, C. M. (2020). Sanctions and tourism: Effects, complexities and research. *Tourism Geographies*, *22*(4–5). doi:10.1080/14616688.2019.1663911

Shafiee, S., Rajabzadeh Ghatari, A., Hasanzadeh, A., & Jahanyan, S. (2021). Smart tourism destinations: A systematic review. *Tourism Review*, *76*(3), 505–528. doi:10.1108/TR-06-2019-0235

Sharifi, A., Khavarian-Garmsir, A. R., & Kummitha, R. K. R. (2021). Contributions of smart city solutions and technologies to resilience against the covid-19 pandemic: A literature review. *Sustainability (Switzerland)*, *13*(14). doi:10.3390/su13148018

Shearmur, R., & Doloreux, D. (2021). The geography of knowledge revisited: Geographies of KIBS use by a new rural industry. *Regional Studies*, *55*(3). doi:10.1080/00343404.2020.1800628

Shelton, T., Zook, M., & Wiig, A. (2015). The "actually existing smart city". *Cambridge Journal of Regions, Economy and Society*, *8*(1), 13–25. doi:10.1093/cjres/rsu026

Silva, B. N., Khan, M., & Han, K. (2018). Towards sustainable smart cities: A review of trends, architectures, components, and open challenges in smart cities. *Sustainable Cities and Society*, *38*(January), 697–713. doi:10.1016/j.scs.2018.01.053

Simonofski, A., Hertoghe, E., Steegmans, M., Snoeck, M., & Wautelet, Y. (2021). Engaging citizens in the smart city through participation platforms: A framework for public servants and developers. *Computers in Human Behavior*, *124*(June), 106901. doi:10.1016/j.chb.2021.106901

Sinaeepourfard, A., Garcia, J., Masip-Bruin, X., Marin-Tordera, E., Cirera, J., Grau, G., & Casaus, F. (2016). Estimating Smart City sensors data generation. In *2016 Mediterranean Ad Hoc Networking Workshop, Med-Hoc-Net 2016 – 15th IFIP MEDHOCNET 2016*. doi:10.1109/MedHocNet.2016.7528424

Sinkiene, J., & Kromolcas, S. (2010). Concept, direction and practice of city attractiveness improvement. *Public Policy and Administration*, *31*(1).

Snieska, V., & Zykiene, I. (2015). City attractiveness for investment: Characteristics and underlying factors. *Procedia – Social and Behavioral Sciences*. doi:10.1016/j.sbspro.2015.11.402

Soares, J. C., Domareski Ruiz, T. C., & Ivars Baidal, J. A. (2021). Smart destinations: A new planning and management approach? *Current Issues in Tourism*. doi:10.1080/13683500.2021.1991897

Sofronijevic, A., Milicevic, V., & Ilic, B. (2014). Smart City as framework for creating competitive advantages in international business management. *Management – Journal for Theory and Practice of Management*, *19*(71), 5–16. doi:10.7595/management.fon.2014.0015

Song, Z., Cao, M., Han, T., & Hickman, R. (2019). Public transport accessibility and housing value uplift: Evidence from the Docklands light railway in London. *Case Studies on Transport Policy*, *7*(3). doi:10.1016/j.cstp.2019.07.001

Sorenson, O., Rivkin, J. W., & Fleming, L. (2006). Complexity, networks and knowledge flow. *Research Policy*. doi:10.1016/j.respol.2006.05.002

Spicer, Z., Goodman, N., & Olmstead, N. (2021). The frontier of digital opportunity: Smart city implementation in small, rural and remote communities in Canada. *Urban Studies*, *58*(3). doi:10.1177/0042098019863666

Staletić, N., Labus, A., Bogdanović, Z., Despotović-Zrakić, M., & Radenković, B. (2020). Citizens' readiness to crowdsource smart city services: A developing country perspective. *Cities*, *107*(June 2019), 102883. doi:10.1016/j.cities.2020.102883

Stam, E., & van de Ven, A. (2021). Entrepreneurial ecosystem elements. *Small Business Economics*, *56*(2). doi:10.1007/s11187-019-00270-6

Stock, W. G. (2011). Informational cities: Analysis and construction of cities in the knowledge society. *Journal of the American Society for Information Science and Technology*. doi:10.1002/asi.21506

Supangkat, S. H., Arman, A. A., Nugraha, R. A., & Fatimah, Y. A. (2018). The implementation of Garuda Smart City Framework for Smart City readiness mapping in Indonesia. *Journal of Asia-Pacific Studies*, *32*(4), 169–176. Retrieved from https://core.ac.uk/download/pdf/159504667.pdf

Tachizawa, E. M., Alvarez-Gil, M. J., & Montes-Sancho, M. J. (2015). How "smart cities" will change supply chain management. *Supply Chain Management*, *20*(3), 237–248. doi:10.1108/SCM-03-2014-0108

Tan, M. (1999). Creating the digital economy: Strategies and perspectives from Singapore. *International Journal of Electronic Commerce*. doi:10.1080/10864415.1999.11518344

Tang, Z., Jayakar, K., Feng, X., Zhang, H., & Peng, R. X. (2019). Identifying smart city archetypes from the bottom up: A content analysis of municipal plans. *Telecommunications Policy*, *43*(10), 101834. doi:10.1016/j.telpol.2019.101834

Tan, S. Y., & Taeihagh, A. (2020). Smart city governance in developing countries: A systematic literature review. *Sustainability (Switzerland)*. doi:10.3390/su12030899

Taylor Buck, N., & While, A. (2017). Competitive urbanism and the limits to smart city innovation: The UK Future Cities initiative. *Urban Studies*, *54*(2), 501–519. doi:10.1177/0042098015597162

Thatcher, J., Burns, R., & Dalton, C. (2021). Introduction to regional approaches to data, environment and society. *Regional Studies*, *55*(12), 1853–1856. doi:10.1080/00343404.2021.1993170

Tiwana, A. (2014). *Platform ecosystems aligning architecture, governance, and strategy*. Library of Congress Cataloging-in-Publication Data.

Trencher, G. (2019). Towards the smart city 2.0: Empirical evidence of using smartness as a tool for tackling social challenges. *Technological Forecasting and Social Change*, *142*(2018), 117–128. doi:10.1016/j.techfore.2018.07.033

Tura, N., & Ojanen, V. (2022). Sustainability-oriented innovations in smart cities: A systematic review and emerging themes. *Cities, 126*(April), 103716. doi:10.1016/j.cities.2022.103716

Ullah, F., Qayyum, S., Thaheem, M. J., Al-Turjman, F., & Sepasgozar, S. M. E. (2021). Risk management in sustainable smart cities governance: A TOE framework. *Technological Forecasting and Social Change, 167*(November 2020). doi:10.1016/j.techfore.2021.120743

UN Habitat. (2008). *The state of the world's cities report*. Nairobi: UN Habitat.

Uşaklı, A., Koç, B., & Sönmez, S. (2017). How 'social' are destinations? Examining European DMO social media usage. *Journal of Destination Marketing & Management, 6*(2), 136–149. doi:10.1016/j.jdmm.2017.02.001

Valdez, A. M., Cook, M., & Potter, S. (2018). Roadmaps to utopia: Tales of the smart city. *Urban Studies, 55*(15), 3385–3403. doi:10.1177/0042098017747857

Vanolo, A. (2014). Smartmentality: The smart city as disciplinary strategy. *Urban Studies, 51*(5), 883–898. doi:10.1177/0042098013494427

Vázquez, J. L., Lanero, A., Gutiérrez, P., & Sahelices, C. (2018). *The contribution of smart cities to quality of life from the view of citizens* (pp. 55–66). doi:10.1007/978-3-319-71014-3_3

Del Vecchio, P., Secundo, G., Maruccia, Y., & Passiante, G. (2019). A system dynamic approach for the smart mobility of people: Implications in the age of big data. *Technological Forecasting and Social Change, 149*(July), 119771. doi:10.1016/j.techfore.2019.119771

Verginer, L., & Riccaboni, M. (2021). Talent goes to global cities: The world network of scientists' mobility. *Research Policy, 50*(1), 104127. doi:10.1016/j.respol.2020.104127

Versini, P. A., Gires, A., Tchiguirinskaia, I., & Schertzer, D. (2020). Fractal analysis of green roof spatial implementation in European cities. *Urban Forestry and Urban Greening, 49*. doi:10.1016/j.ufug.2020.126629

Vial, G. (2019). Understanding digital transformation: A review and a research agenda. *The Journal of Strategic Information Systems, 28*(2), 118–144. doi:10.1016/j.jsis.2019.01.003

Vidiasova, L., & Cronemberger, F. (2020). Discrepancies in perceptions of smart city initiatives in Saint Petersburg, Russia. *Sustainable Cities and Society*. doi:10.1016/j.scs.2020.102158

Vinod Kumar, T. M. (2020). Smart environment for smart cities. In *Advances in 21st century human settlements*. doi:10.1007/978-981-13-6822-6_1

Vith, S., Oberg, A., Höllerer, M. A., & Meyer, R. E. (2019). Envisioning the 'Sharing City': Governance strategies for the sharing economy. *Journal of Business Ethics, 159*(4), 1023–1046. doi:10.1007/s10551-019-04242-4

Vu, K., & Hartley, K. (2018). Promoting smart cities in developing countries: Policy insights from Vietnam. *Telecommunications Policy, 42*(10), 845–859. doi:10.1016/j.telpol.2017.10.005

Wang, D., Li, X., & Li, Y. (2013). China's "smart tourism destination" initiative: A taste of the service-dominant logic. *Journal of Destination Marketing & Management, 2*(2). doi:10.1016/j.jdmm.2013.05.004

Wang, A., Wang, P., Miao, X., Li, X., Ye, N., & Liu, Y. (2020). A review on non-terrestrial wireless technologies for Smart City Internet of Things. *International Journal of Distributed Sensor Networks, 16*. doi:10.1177/1550147720936824

Wang, J., Xie, C., Huang, Q., & Morrison, A. M. (2020). Smart tourism destination experiences: The mediating impact of arousal levels. *Tourism Management Perspectives*, *35*(December 2019), 100707. doi:10.1016/j.tmp.2020.100707

Werner, S. (2002). Recent developments in international management research: A review of 20 Top Management Journals. *Journal of Management*. doi:10.1177/014920630202800303

Wicaksana, G. B. A. (2020). Future city based on smart mobility concept: Character and benchmarking. *Journal of Architectural Research and Education*, *2*(1), 10. doi:10.17509/jare.v2i1.24112

Wiig, A. (2015). IBM's smart city as techno-utopian policy mobility. *City*. doi:10.1080/13604813.2015.1016275

Wiig, A. (2016). The empty rhetoric of the smart city: From digital inclusion to economic promotion in Philadelphia. *Urban Geography*, *37*(4), 535–553. doi:10.1080/02723638.2015.1065686

Windmeijer, F. (2005). A finite sample correction for the variance of linear efficient two-step GMM estimators. *Journal of Econometrics*, *126*(1), 25–51. doi:10.1016/j.jeconom.2004.02.005

Winters, J. V. (2011). Why are smart cities growing? who moves and who stays. *Journal of Regional Science*, *51*(2), 253–270. doi:10.1111/j.1467-9787.2010.00693.x

Wooldridge, J. M. (2002). *Econometric analysis of cross section and panel data* (Vol. 58, No. 2). doi:10.1515/humr.2003.021

Xiahou, X., Yuan, J., Xie, H., Skibniewski, M. J., & Li, Q. (2020). Exploring driving factors of smart city development under the physical-human society-cyber (P-H-C) space model. *International Journal of Construction Management*, *0*(0), 1–11. doi:10.1080/15623599.2020.1824601

Xie, B., Zhou, J., & Luo, X. (2016). Mapping spatial variation of population aging in China's mega cities. *Journal of Maps*, *12*(1), 181–192. doi:10.1080/17445647.2014.1000984

Yang, C. (2020). Historicizing the smart cities: Genealogy as a method of critique for smart urbanism. *Telematics and Informatics*, *55*(April), 101438. doi:10.1016/j.tele.2020.101438

Yan, J., Liu, J., & Tseng, F. M. (2020). An evaluation system based on the self-organizing system framework of smart cities: A case study of smart transportation systems in China. *Technological Forecasting and Social Change*, *153*(1), 119371. doi:10.1016/j.techfore.2018.07.009

Yigitcanlar, T., Han, H., Kamruzzaman, M., Ioppolo, G., & Sabatini-Marques, J. (2019). The making of smart cities: Are Songdo, Masdar, Amsterdam, San Francisco and Brisbane the best we could build? *Land Use Policy*, *88*. doi:10.1016/j.landusepol.2019.104187

Yigitcanlar, T., Kankanamge, N., & Vella, K. (2020). How are smart city concepts and technologies perceived and utilized? A systematic Geo-Twitter analysis of smart cities in Australia. *Journal of Urban Technology*. doi:10.1080/10630732.2020.1753483

Yuan, B., & Li, J. (2019). The policy effect of the general data protection regulation (GDPR) on the digital public health sector in the European Union: An empirical investigation. *International Journal of Environmental Research and Public Health*, *16*(6). doi:10.3390/ijerph16061070

Yuan, Y., Lu, Y., Chow, T. E., Ye, C., Alyaqout, A., & Liu, Y. (2020). The missing parts from social media–enabled smart cities: Who, where, when, and what? *Annals of the Association of American Geographers, 110*(2). doi:10.1080/24694452.2019.1631144

Yun, Y., & Lee, M. (2019). Smart City 4.0 from the perspective of open innovation. *Journal of Open Innovation: Technology, Market, and Complexity, 5*(4). doi: 10.3390/joitmc5040092

Zavattaro, S. M., & Sementelli, A. J. (2014). A critical examination of social media adoption in government: Introducing omnipresence. *Government Information Quarterly, 31*(2), 257–264. doi:10.1016/j.giq.2013.10.007

Zhang, N., Zhao, X., & He, X. (2020). Understanding the relationships between information architectures and business models: An empirical study on the success configurations of smart communities. *Government Information Quarterly, 37*(2), 101439. doi:10.1016/j.giq.2019.101439

Zhao, F., Fashola, O. I., Olarewaju, T. I., & Onwumere, I. (2021). Smart city research: A holistic and state-of-the-art literature review. *Cities, 119*(May 2020), 103406. doi: 10.1016/j.cities.2021.103406

Zhu, S., Li, D., & Feng, H. (2019). Is smart city resilient? Evidence from China. *Sustainable Cities and Society, 50*. doi:10.1016/j.scs.2019.101636

Zygiaris, S. (2013). Smart city reference model: Assisting planners to conceptualize the building of smart city innovation ecosystems. *Journal of the Knowledge Economy, 4*(2). doi:10.1007/s13132-012-0089-4

Index

Printed and bound by CPI Group (UK) Ltd, Croydon, CR0 4YY
18/09/2026
14965017-0001